Ormsby MacKnight Mitchel

The Planetary and Stellar Worlds

Ormsby MacKnight Mitchel
The Planetary and Stellar Worlds
ISBN/EAN: 9783337371555

Printed in Europe, USA, Canada, Australia, Japan

Cover: Foto ©berggeist007 / pixelio.de

More available books at **www.hansebooks.com**

THE
PLANETARY

AND

STELLAR WORLDS.

A POPULAR EXPOSITION OF THE GREAT DISCOVERIES AND
THEORIES OF MODERN ASTRONOMY.

BY

GEN. O. M. MITCHEL.

NEW YORK:
JOHN B. ALDEN, PUBLISHER.
1887.

TABLE OF CONTENTS.

LECTURE I.
An Exposition of the Problem which the Heavens Present for Solution.

First revolution of the Heavens witnessed by man, 11
The curiosity excited and its effects,............. 12
Object of the course,.................... 12
The astronomer lives in all ages,................. 13
First problem to distinguish between real and apparent motion,........................... 14
The relations of the sun, and earth, and moon,... 15
The wandering stars or planets,................. 15
Fixed stars, points of reference,..... 16
Complexity of the planetary movements,..... ... 16
Discovery of the true centre of motion,........... 17

The planets form a system; their orbits, laws, stability,..... 18
Perturbations,........ 19
Analytic machinery,........................... 19
Examination of the starry heavens...... 20
Distance of the fixed stars to be determined,..... 20
Motions among the stars,...................... 21
Binary systems, movement of the sun in space,.... 21
Investigations yet to be made, and probable success,................................. 22

LECTURE II.
The Discovery of the Primitive Ages.

The founder of the science of astronomy unknown, 25
First discovery on the moon, ... 26
Her motion among the stars and her phases,...... 26
Cause of the phases sought; two revolutions of the moon discovered,........................... 27
First ideas of the constellations; North Star,. ... 28
Motions of the sun and moon among the stars; the starry heavens surround the earth,........ 28
First measure of the year,........ 29
A moving star discovered,...................... 29
Periods of the planets determined,................ 31
The sun's apparent motion the subject of perplexity,.... 32

The equinoxes and solstices,.... 33
Inequality in the sun's motion detected,.......... 34
The construction of the sphere,......... 34
Its uses in observing,........................... 34
Eclipses of the sun and moon, and their effects,... 35
Explanation of solar eclipses; discovery of the moon's reflective light, and explanation of her phases, ... 35
Lunar eclipses explained,.................. 36
Prediction of the first eclipse,...................... 38
Value of recorded eclipses, ... 39

LECTURE III.
Theories for the Explanation of the Motions of the Heavenly Bodies.

The names of early discoverers lost,............. 41
Chaldean period found among many nations,..... 41
The days of the week, and their names; place of the vernal equinox,........................ 42
Precession of the equinoxes, ... 43
Astronomy of the primitive ages, and condition of the mind at the beginning of the era reached by history,................................ 44
Causes retarding the progress of astronomy,..... 45
The confounding of true and apparent motion; immobility of the earth and its central position, 45
The Greek astronomers, Pythagoras, Nicetas, Hipparchus, Ptolemy, ... 46

System of Ptolemy, ... 48
Astronomy cultivated by the Arabs,............. 48
The era of modern science commences, ... 49
Copernicus and his discoveries,..................... 49
His system promulged with great caution,........ 51
Kepler, the character of his mind, his mode of research,.... 52
His great discoveries; he finds the orbits of the planets,................................... 53
Detects his second law,.................... 54
His efforts to find the third law successful,........ 57
Importance of these laws,........................ 57

TABLE OF CONTENTS.

LECTURE IV.
Discovery of the Great Laws of Motion and Gravitation.

The philosophy of Aristotle—its hold on the mind, 59	Problem presented for solution to Newton, ... 67
Characteristics of Galileo's mind, ... 59	Conjectures of Kepler; discoveries of Descartes, .. 67
He detects the errors of Aristotle, ... 60	Measure of the moon's distance and of the earth's
Attacks his theories, and demonstrates their false-	circumference, ... 68
hood by experiment, ... 60	The law of gravitation—Newton's first effort to
Driven from Pisa by his enemies, ... 61	demonstrate the truth of this law, ... 69
Discovers the law of falling bodies, ... 61	He finally shows that the moon is ever falling
Adopts the Copernican theory, ... 62	towards the earth, and proves the law, ... 71
Constructs a telescope, ... 62	Enunciation of his great law, ... 72
His discoveries in the moon, and among the plan-	Discovers that the planets may revolve in conic
ets and fixed stars, ... 63	sections, ... 73
Phases of Venus, ... 64	Contrast between Kepler's and Newton's methods
Questions relating to the planetary motions, ... 65	of research, ... 73
Laws of motion; the centrifugal force, ... 66	

LECTURE V.
Universal Gravitation Applied to the Explanation of the Phenomena of the Solar System.

The era of physical astronomy commences, ... 75	Perplexity occasioned by the seeming discrepancy
A theoretic system proposed and discussed, a	between the observed and computed motion of
central sun and solitary planet, ... 76	the moon's perigee, ... 84
Planets and satellites added and their effects con-	Finally removed by Clairault, ... 85
sidered, ... 77	Changes in the earth's figure occasioned by its
How the imagined system may be made the sys-	rotation, ... 86
tem of nature, ... 78	The form of equilibrium reached, ... 88
Discussion of the relative motions of the sun,	The precession of the equinoxes caused by the
moon, and earth, under the action of their	protuberant matter at the earth's equator, ... 89
mutual influences, ... 78	Moon affected by the redundant matter at the
The moon's acceleration; motion of her apsides,	earth's equator, ... 91
nodes, etc., and the discovery of the change	Wonderful questions answered by an examination
in the figure of the earth's orbit, ... 81	of the moon, ... 91

LECTURE VI.
The Stability of the Planetary System.

Rapid survey of the system, ... 93	Stability of the principal axes, ... 100
General characteristics of the planets, ... 94	Motion of the perihelion, ... 100
What phenomena gravitation must account for, .. 95	The inclinations, ... 102
Stability not the sole object of the Creator, ... 95	The lines of nodes, ... 103
Laws of matter selected in wisdom, ... 96	The periodic times, ... 103
By how much does the central force diminish the	Stability of the great system, ... 104
primitive velocity of the planet? 97	Of the system of the earth and moon, ... 105
Changes in the elements of the orbits of the	Of Jupiter's system, ... 106
planets, ... 98	Of Saturn's system, ... 106
The eccentricity, ... 100	

LECTURE VII.
The Discovery of the New Planets.

Kepler's speculations, ... 109	The symmetry of the system destroyed by the dis-
Discovery of Uranus, ... 110	covery of Pallas, ... 114
Bode's law of interplanetary spaces, ... 110	Olbers's theory of the bursting of a planet, ... 114
The astronomical congress of Lilienthal, in 1800, .. 111	Discovery of Juno and Vesta, ... 114
Piazzi's discovery of a new planet, Ceres; its loss	Hencke discovers Astrea and Hebe, ... 115
and rediscovery, ... 112	Hind discovers Iris and Flora, ... 115

TABLE OF CONTENTS.

Lecture VII., The Discovery of the New Planets.—*Continued.*

Search for a planet beyond Uranus commenced,..116
Causes of this search,................................116
Leverrier's researches on Mercury,............117
Its transit in May, 1845, ...118
Leverrier presents his computations to the French Academy,...119
Popular exhibition of his reasoning,...........119

The hypothetical planet found by Galle, of Berlin,121
Adams's computations, ...122
The new planet detected by its disc, ...122
Walker's computations,..............................123
Pierce's views,...124
Leverrier claims Neptune to be the planet of theory,..124

LECTURE VIII.
The Cometary Worlds.

Characteristics of comets,.........................125
Reduced to law by Newton,.......................126
The comet of 1680,..................................126
Halley's comet of 1682,............................126
Its return in 1759 predicted,.....................128
Its return in 1835,...................................129
Wonderful changes in its magnitude, ...130
Encke's comet,..131
Approaching the sun,...............................131
Resisting medium,...................................133
Biela's comet,...133
Fears excited of collision with the earth in 1832,...134

Its nebulous character, ...134
Its double character in 1846. Separation of the comets,..135
Vast periods of some comets,....................136
Comets seem to transit the sun's disc,......137
Comets accounted for by Laplace's nebular hypothesis,..138
Herschel's theory of the physical condition of comets,..139
His theory accounts for the diminishing period of Encke's comets, ...140
Zodiacal light, ...140

LECTURE IX.
The Scale on Which the Universe is Built.

Scale of the planetary system,..................143
Radius of the earth's orbit too small a unit,.....143
The velocity of light determined from the eclipses of Jupiter's satellites, and employed as a unit,144
Parallax of the fixed stars,.......................145
No parallax sensible to the naked eye,.......146
Great distance of the fixed stars inferred from this fact,..146

Bradley's researches for parallax,...............148
Discovery of nutation and its value,...........149
Discovery of aberration—its explanation,....150
Herschel's researches for parallax,............151
Discovery of the revolving stars, ...152
Power of modern telescopes,....................153
Bessel discovers the parallax of 61 Cygni,....154

LECTURE X.
The Motions and Revolutions of the Fixed Stars.

Distances separating man from the stars,...163
Various difficulties in the research for their motions,...163
Hipparchus discovers a new and brilliant star,....164
The new star of 1572,.............................165
The new star of 1604,.............................165
The disappearance of old stars, ...166
Changes of Algol,...................................166
Periodical stars,.....................................166
Gravitation extended to the sphere of the fixed stars, ...167
Periods of some of the binary systems,......168
Herschel sounds the depth of the Milky Way,.....169

Seeks the direction of the solar motion,.....169
His reasoning,.......................................170
Argelander's research for the point towards which the solar system is moving,..................172
Struve's investigation for the quantity of angular motion of the system, as seen from stars of the first magnitude,..................................173
His father's research for the relative distances of stars of different magnitudes,................174
Peters's research for the parallax of stars of the second magnitude,...............................175
Maedler's theory of the central sun,..........179
The attributes of God as displayed in the universe,182

PREFACE.

A FEW words in explanation of the circumstances under which this volume is presented to the public may not be unacceptable to the reader. It is now a little more than six years since the writer conceived the idea of erecting a great astronomical observatory in the city of Cincinnati. My attention had been for many years directed to this subject, by the duties of the professorship, which I then held in the college. In attempting to communicate the great truths of astronomy, there were no instruments at hand, to confirm and fix the wonderful facts recorded in the books. Up to that period our country, and the West particularly had given but little attention to practical astronomy. A few individuals, with a zeal and ardor deserving of all praise, had struggled on to eminence almost without means or instruments. An isolated telescope was found here and there scattered through the country; but no regularly organized observatory with powerful instruments existed within the limits of the United States, so far as I know.

To attempt the building of an observatory of the first class, and to furnish it with instruments of the highest order, without any aid from the general or state government, but by the voluntary contribution of all classes of citizens, was an enterprise of no common difficulty. To ascertain whether any interest could be excited in the public mind in favor of astronomy, in the spring of 1842 a series of lectures was delivered in the hall of the Cincinnati College. To give an increased effect to these discourses (which were unwritten, and in a style of great simplicity), a mechanical contrivance was prepared, by the aid of which the beautiful telescopic views in the heavens were presented to the audience, with a brilliancy and power scarcely inferior to that displayed by the most powerful telescopes. To this fortunate invention were these lectures, no doubt, principally indebted for the interest which they produced and which occasioned them to be attended by a very large number of the intelligent persons in the city. Encouraged by the large audiences, which continued through two months to fill the lecture-room, and still more by the request to repeat the last lecture of the course in one of the great churches of the city, I matured a plan for the building of an observatory, which it was resolved should be presented to the audience at the close of the lecture in case circumstances should favor. Through the kindness of a few friends who were now beginning to take a deep interest in the matter, more than two thousand persons were in attendance; and it seemed that the moment had arrived for taking the first step in an enterprise whose fate it was impossible to predict.

Having closed the subject under discussion, the audience were requested to give me a few minutes of time, for the explanation of a matter which it was hoped would not be received without some feelings of interest and approbation. The rapid advances of astronomy in Europe were then referred to—the rection of observatories in all parts of the world—the variety of magnificent instruments in Russia and Germany, in France and England, and the utter deficiency of our own country in everything pretaining to the science of the stars. The past neglect was easily accounted for, and might be excused; the future scientific character of the country rested with the people, and upon them devolved the responsibility of providing the means for original research. In Europe, imperial treasure and princely munificence could build the temples of science; under a free government no such means existed, and to accomplish the erection of these great scientific institutions, the intelligent liberality of the whole community was the only resource. But it had been denied that this resource could be relied on; and it had been roundly asserted that, in the nature of things, the United States must ever remain grossly defective in all the appliances for scientific research. To test the truth or falsehood of these statements was not a difficult matter; and thus encouraged by the interest already manifested in behalf of astronomy, I had already resolved to devote *five* years of faithful effort to accomplish the erection of a great astronomical observatory in the city of Cincinnati.

PREFACE.

This announcement was received with every mark of favor, and the following simple plan was at one presented. The entire amount required to erect the buildings and purchase the instruments, should be divided into shares of twenty-five dollars; every shareholder to be entitled to the privileges of the observatory, under the management of a board of control, to be elected by the shareholders. Before any subscription should become binding, the names of three hundred subscribers should be first obtained. This accomplished, these three hundred should meet, organize, and elect a board, who should thenceforward manage the affairs of the association.

Such is the history of the Cincinnati Astronomical Society. Two resolutions were taken in the outset, to which I am indebted for any success which may have attended my own personal efforts. First: To work faithfully for *five years*, during all the leisure which could be spared from my regular duties. Second : Never to become *angry*, under any provocation, while in the prosecution of this enterprise.

In three weeks the three hundred subscribers had been obtained. No public meeting had been called; and these names had been procured by private solicitation, and a personal explanation of the nature and advantages of the enterprise. So soon as the number was complete, the subscribers convened, organized, elected officers and a directory, and gave me a commission to visit Europe, to procure instruments, examine observatories, and obtain the requisite knowledge to erect and conduct the institution which it was now hoped would be one day reared.

This order being received, on the second day I started for New York, and on the 16th of June, 1842, sailed for Liverpool. Having visited many of the best appointed observatories both in England and on the Continent (in each and every one of which I was received with a degree of kindness and attention for which I acknowledge the deepest obligations), and having been unsuccessful in finding, either in London or Paris, an object-glass of the size required, I finally determined to visit the city of Munich. The fame of the optical institute of the celebrated Frauenhofer had even reached the banks of the Ohio; and it was hoped that, in that great manufactory, an instrument such as the society desired might be obtained, if not completed, at least in such a state of forwardness as to permit it to be furnished at an early day. In this I was not disappointed. An object-glass of nearly twelve inches diameter, and of superior finish, was found in the cabinet of Mr. Mertz, the successor of Frauenhofer. This glass had been subjected to a severe trial in the tube of the great refractor of the Munich observatory, by Dr. Lamont, and had been pronounced of the highest quality.

To mount this glass would require about two years, at a cost of nearly ten thousand dollars; a sum considerably greater than that appropriated at the time for an equatorial telescope. Having made a conditional arrangement for this and other instruments, I returned to Greenwich, England, where, at the invitation of Professor Airy, the Astronomer Royal, I remained for some time to study. Having accomplished the objects of my journey, I returned home, and rendered a report to a very large meeting of the members of the association and other citizens of Cincinnati.

During my absence of four months, a great change had occurred in the commercial affairs of the country. Everything was depressed to the lowest point, and increased in a high ratio the necessary difficulties of such an undertaking, always great, even if carried forward at a time when the country is prosperous.

With great difficulty the subscription was increased to an amount sufficient to warrant the ordering of the great object-glass already referred to. The sum of three thousand dollars was collected and remitted to meet the first payment. Even this fraction of the entire sum was collected with difficulty; but as the remaining part of the price of the telescope was not to be paid until the completion of the instrument, it was hoped that the ample time thus allowed would render the task of collection comparatively easy.

The principal instrument having been ordered, and the first payment of its cost made, attention was now given to the procuring of a suitable site for the building. Fortunately for the society, the place of all others most perfectly adapted to their wants was then the property of Nicholas Longworth, Esq. It is a lofty hill-top, rising some four hundred feet above the level of the city, and commanding a perfect horizon in all directions. On making known to Mr. Longworth the prospects and wants of the Astronomical Society, the writer was directed by him to select *four acres* on the hill-top, out of a tract of some twenty-five acres, and to proceed at once to enclose it, as it would give him great pleasure to present it to the association. On compliance with the conditions of the title-bond, a deed has since been received, placing the society in full possession of this elegant position.

Preparations were now made to commence the erection of the building for the observatory. The grounds were inclosed, a road built, rendering the access to the hill-top comparatively easy, the excavation for the foundations were made, and, on the 9th day of November, 1843, the corner-stone of the pier which was to sustain the great Refracting Telescope, was laid by *John Quincy Adams*, with appropriate ceremonies. On this occasion Mr. Adams made his last great oration. The deep interest which he had taken in astronomical science warranted the hope that he might be induced to visit the West, on the occasion of laying the foundation-stone of the first great popular observatory ever erected in the United States. This hope was not disappointed. The unaffected devotion of this truly great man to the interests of his country were, perhaps, never more perfectly exhibited than in his ready acquiescence to comply with the wishes of the astronomical society, that he should perform for them the important services on which the future success of this new enterprise in no small degree depended. His high character, his advanced age, the length of the journey, the inclemency of the season, all combined to exhibit to his countrymen the depth of his interest in a cause which could induce such sacrifices.

After the laying of the corner-stone, the lateness of the season and other causes induced a suspension of the work on the building for the winter; and it was not resumed until May, 1844. In the mean time, after incredible difficulty, the entire amount called for in the payment for the great telescope was collected and remitted; and the society was left with scarcely a dollar of available means to commence the erection of a building which, according to the plan, would cost some seven or eight thousand dollars.

It was believed that the intelligent mechanics of Cincinnati would lend their powerful aid in the accomplishment of an enterprise which had progressed far enough to give some confidence in its ultimate success. With little or no means the building was commenced, trusting to activity and perseverance to supply the means as the work progressed. During the first week but three workmen were employed; but by the commencement of the next week the funds had been obtained to pay these, and to double their number. In six weeks not less than one hundred hands were at work on the hill-top and in the city. Mechanics of all trades subscribed for stock, and paid their subscription in work. The stone of which the building is erected was quarried from the grounds of the society. The lime was burnt on the hill, and every means was adopted to reduce the necessary expenditure. Payment for stock in the society was received in every possible article of trade; due-bills were taken, and these were converted into others which would serve in the payment of bills.

In this way the building was reared, and finally covered in, without incurring any debt. But the conditions of the bonds by which the lot of ground was held required the completion of the observatory in two years from its date; and these two years would expire June, 1845. It was seen to be impossible to carry forward the building fast enough to secure its completion by the required time, without incurring some debt. My own private resources were used, in the hope that a short time after the finishing of the observatory would be sufficient to furnish the funds to meet all engagements. The work was pushed rapidly forward. In February, 1845, the great telescope safely reached the city of Cincinnati; and in March the building was ready for its reception. I had now exhausted all my private means, and to increase the difficulty of the position in which I was placed, the college edifice took fire and burned to the ground. My ordinary means of support were thus destroyed at a single blow. I had engaged to conduct the observatory, without compensation from the society, for ten years, in the hope that my college salary would be sufficient for my wants. It was impossible to abandon the observatory. The college could not be rebuilt, at least for several years, and in this emergency I found it necessary to seek some means of support, least inconsistent with my duties in the observatory. My public lectures at home had been comparatively well received, and after much hesitation it was resolved to make an experiment elsewhere. For five years I had been pleading the cause of science among those little acquainted with its technical language. I had become habituated to the use of such terms as were easily understood; and probably to this circumstance, more than to any other one thing, am I indebted for any success which may have attended my public lectures. To the citizens of Boston, Brooklyn, New York, and New Orleans, for the kindness with which they were pleased to receive my imperfect efforts, I am deeply indebted. My lectures were never written, and no idea was entertained of publishing a course, until the partiality of my friends induced me to attempt this experiment.

Such are the circumstances under which this effort to trace the career of the human mind, in its researches among the stars, has been undertaken. No one science, perhaps, so

PREFACE.

perfectly illustrates the gradual growth and development of the powers of human genius. The movement of the mind has been constantly onward—its highest energies have ever been called into requisition—and there never has been a time when astronomy did not present problems not only equal to all that man could do, but passing beyond the limits of his greatest intellectual vigor. Hence, in all ages and countries, the absolute strength of human genius may be measured by its reach to unfold the mysteries of the stars.

It will be seen that in the following lectures one single object has engaged the attention of the writer,—*the structure of the universe, so far as revealed by the mind of man.* The uses of science have in no way been considered. The effects on the mind, on society, on civilization, on commerce, on religion, have not been permitted to mar the unity of the original design. The onward, steady, triumphant march of mind, in its study and exploration of the universe of God has been my only object, the single theme of the entire series.

CINCINNATI OBSERVATORY,
Mount Adams, *May*, 1848.

THE STRUCTURE OF THE UNIVERSE

INTRODUCTORY LECTURE.

AN EXPOSITION OF THE PROBLEM WHICH THE HEAVENS PRESENT FOR SOLUTION.

THE subject to which your attention is invited, claims no specific connection with the every-day struggle of human life. Far away from the earth on which we dwell, in the blue ocean of space, thousands of bright orbs, in clusterings and configurations of exceeding beauty, invite the upward gaze of man, and tempt him to the examination of the wonderful sphere by which he is surrounded. The starry heavens do not display their glittering constellations in the glare of day, while the rush and turmoil of business incapacitate man for the enjoyment of their solemn grandeur. It is in the stillness of the midnight hour, when all nature is hushed in repose, when the hum of the world's on-going is no longer heard, that the planets roll and shine, and the bright stars, trooping through the deep heavens, speak to the willing spirit that would learn their mysterious being.

Often have I swept backward in imagination six thousand years, and stood beside our Great Ancestor, as he gazed for the first time upon the going down of the sun. What strange sensations must have swept through his bewildered mind, as he watched the last departing ray of the sinking orb, unconscious whether he should ever behold its return. Wrapt in a maze of thought, strange and startling, his eye long lingers about the point at which the sun had slowly faded from his view. A mysterious darkness, hitherto unexperienced, creeps over the face of nature. The beautiful scenes of earth, which through the swift hours of the first wonderful day of his existence, had so charmed his senses, are slowly fading one by one from his dimmed vision. A gloom deeper than that which covers earth, steals across the mind of earth's solitary inhabitant. He raises his inquiring gaze towards heaven, and lo! a silver crescent of light, clear and beautiful, hanging in the western sky, meets his astonished eye. The young moon charms his untutored vision, and leads him upward to her bright attendants, which are now stealing one by one, from out the deep blue sky.

The solitary gazer bows, and wonders, and adores. The hours glide by—the silver moon is gone— the stars are rising—slowly ascending the heights of heaven—and solemnly sweeping downward in the stillness of the night. The first grand revolution to mortal vision is nearly completed. A faint streak of rosy light is seen in the east—it brightens—the stars fade—the planets are extinguished—the eye is fixed in mute astonishment on the growing splendor, till the first rays of the returning sun dart their radiance on the young earth and its solitary inhabitant. To him "the evening and the morning were the first day."

The curiosity excited on this first sole nn night—the consciouness that in the heavens God had declared his glory -the eager desire to comprehend the mysteries that dwell in these bright orbs, have clung to the descendants of him who first watched and wondered, through the long lapse of six thousand years. In this boundless field of investigation, human genius has won its most signal victories—Generation after generation has rolled away, age after age has swept silently by, but each has swelled by its contributions the stream of discovery—One barrier after another has given away to the force of intellect—mysterious movements have been unravelled—mighty laws have been revealed—ponderous orbs have been weighed, their reciprocal influences computed, their complex wanderings made clear, until the mind, majestic in its strength, has mounted step by step up the rocky height of its self-built pyramid, from whose star-crowned summit it looks out upon the grandeur of the universe, self-clothed with the prescience of a God.—With resistless energy it rolls back the tide of time, and lives in the configuration of rolling worlds a thousand years ago, or more wonderful, it sweeps away the dark curtain from the future, and beholds those celestial scenes which shall greet the vision of generations when a thousand years shall have rolled away, breaking their noiseless waves on the dim shores of eternity.

To trace the efforts of the human mind in this long and ardent struggle, to reveal its hopes and fears, its long years of patient watching, its moments of despair and hours of triumph—to develop the means by which the deep foundations of the rock-built pyramid of science have been laid, and to follow it as it slowly rears its stately form from age to age, until its vertex pierces the very heavens—these are the objects, proposed for accomplishment and these are the topics to which I would invite your earnest attention. The task is one of no ordinary difficulty. It is no feast of fancy, with music and poetry, with eloquence and art, to enchain the mind. Music is here—but it is the deep and solemn harmony of the spheres. Poetry is here—but it must be read in the characters of light, written on the sable garments of night. Architecture is here—but it is the colossal structure of sun and system, of cluster and universe. Eloquence is here—but "there is neither speech nor language—Its voice is not heard," yet its resistless sweep comes over us in the mighty periods of revolving worlds.

Shall we not listen to this music, because it is deep and solemn? Shall we not read this poetry, because its letters are the stars of heaven ?—Shall we refuse to contemplate this architecture, because "its architraves, its archways, seem ghostly from infinitude ?" Shall we turn away from this surging eloquence, because its utterance is made through sweeping worlds? No—the mind is ever inquisitive, ever ready to attempt to scale the most rugged steeps. Wake up its enthusiasm—fling the light of hope on its pathway, and no matter how rough and steep and rocky it may prove, *onward!* is the word which charms its willing powers.

It is not my wish or design to introduce you to the dark technicalities of science, neither do I propose to rest satisfied with the barren statement of the results which have been reached by the efforts of genius. While on the one hand I shall endeavor to shun all attempt at critical scientific demonstration, which could only be intelligible to the professed student of astronomy, I shall on the other hand fearlessly attempt such an exposition of the processes and trains of reasoning by which great truths have been elicited, as to show to every intelligent mind that the problem is not impossible; by simplicity of language, by familiar illustrations, to fling light enough upon these mysterious propositions, to show a pathway, though it be dim and rugged, still a pathway, which if pursued shall certainly lead to a full and perfect solution. I ask, then, no critical previous knowledge of the subject, on the part of those who would follow me in the wonderful developments which I am about to attempt. Give me but your earnest and unbroken attention. Go with me in imagination, and join in the nightly vigils of the astronomer, and while his mind with powerful energy struggles with difficulty, join your own sympathetic efforts with his—hope with his hope—tremble with his fears—rejoice with his triumphs. Lend me but this kind of interest, and my task is already half accomplished.

Before proceeding to an actual exposition of the structure of the Heavens, I propose in this introductory lecture to announce the nature of the problem, which the mind has essayed to resolve, and to point out the more important auxiliaries, mental and mechanical, which it has conjured to its aid. If the difficulties of this problem should overwhelm the mind, let it be remembered that the astronomer has ever lived, and never dies. The sentinel upon the watchtower is relieved from duty, but another takes his place, and the vigil is unbroken. No—the astronomer never dies. He commences his investigations on the hill tops of Eden—he studies the stars through the long centuries of antediluvian life. The deluge sweeps from the earth its inhabitants, their cities and their monuments—but when the storm is hushed, and the heavens shine forth in beauty, from the summit of Mount Ararat the astronomer resumes his endless vigils. In Babylon he keeps his watch, and among the Egyptian priests he inspires a thirst for the sacred mysteries of the stars. The plains of Shinar—the temples of India—the pyramids of Egypt, are equally his watching places. When science fled to Greece, his home was in the schools of her philosophers; and when darkness covered the earth for a thousand years, he pursues his never-ending task from amidst the burning deserts of Arabia. When science dawned on Europe, the astronomer was there—toiling with Copernicus—watching with Tycho—suffering with Galileo—triumphing with Kepler. Six thousand years have rolled away since the grand investigation commenced. We stand at the terminus of this vast period, and looking back through the long vista of departed years, mark with honest pride the successive triumphs of our race. Midway between the past and future, we sweep backward and witness the first rude effort to explain the celestial phenomena—we may equally stretch forward thousands of years, and although we cannot comprehend what shall be the condition of astronomical science at that remote period, of one thing we are certain—the past, the present, and the future, constitute but one unbroken chain of observations, condoning all time, to the astronomer, into one mighty *now*.

From the vantage ground which we occupy, it will not be difficult to announce so much of the great problem as has already been resolved, and to form some approximate conception of what remains for future ages to accomplish.

In the exposition about to be attempted, I do not propose to present any trains of reasoning, or any results which may have been reached. These shall engage our attention hereafter. At present permit me simply to translate into language the questions which the visible heavens propound.

The most cursory examination of the celestial vault reveals the fact, that not one solitary object, visible to the eye, is at rest. Motion is the attribute of sun and moon and planets and stars. The earth we inhabit alone remains fixed, to the senses.

The first great problem propounded for human ingenuity, is to sever *real motion* from that which is *unreal* and only *apparent*. To accomplish this, some knowledge of the form of the earth which we inhabit must be obtained. Not only must we acquire a knowledge of its figure, but in like manner we must learn with certainty its actual condition, whether of rest or motion. If at absolute rest in the center of the universe, then the rising sun, the setting moon, the revolving heavens, are real exhibitions, and must be examined as such. On the contrary, should it be found to be imposible to predicate of the earth absolute immobility, then arises the complicated question, how many motions belong to it? and with what velocity does it move? If a motion of rotation exist, what is the position of the axis about which it revolves, and is this axis permanent or changeable? If a motion of translation in space must be adopted, then whither is the earth urging its flight? what the nature of the path described? the velocity of its movement, and the laws by which it is governed? These are some of the questions which present themselves in the outset, touching

INTRODUCTORY LECTURE. 1

the condition of the earth, on whose surface the astronomer is located, in his researches of the heavens.

Beyond the limits of the earth, a multitude of objections present themselves for examination: and first of all the sun, the great source of life and light and heat, demands the attention of the student of the heavens. That some inscrutable tie binds it to the earth, or the earth to it, was early recognized in the fact, that whether the sun was moving or at rest, the relative distance of it and the earth never changed by any great amount; and whatever changes did occur, were all obliterated in a short period, and the distance by which these bodies are separated was restored to its primitive value, to recommence its cycle of changes in the same precise order.—Here then was a grand problem, to determine the relations existing between the sun and earth; to endue with motion that one of these bodies which did move, and to fix the limits within which the observed changes occurred, both in time and distance.

While the connection between the sun and earth was certain, a mutual dependence between the earth and the other great source of light, the moon, was equally manifest. The invariability in the apparent diameter of the moon, demonstrates the fact, that whether the earth were moving or stationary, the moon never parts company with our planet. In all her wanderings among the fixed stars, in her elongations from the sun, in her wondrous phases and perpetual changes, some invisible hand held her at the same absolute distance from the earth. But to decide whether this power resided in the earth or the moon, or in both, to explain these wondrous changes from the silver crescent of the western sky to the full orb which rose with the setting sun, pouring a flood of light over all the earth, to develop the mysterious connection between the disappearance of the moon and those terrific phenomena, the going out of the sun in dim eclipse— these furnished themes for investigation requiring long centuries of patient watching of never-ending toil.

Passing out from the sun and moon to the more distant stars, among the brightest of those which gemmed the nocturnal heavens, a few were found differing from all the rest in the fact that they wandered from point to point, and at the end of intervals widely differing among themselves, swept round the entire heavens, and returned to their starting point, to recommence their ceaseless journies. These were named planets *wanderers*, in contradistinction to the host of stars which were fixed in position, unchanged from century to century.

Hence arose a new and profound series of investigations: where were these wandering stars urging their flight? Were their motions real or apparent? Were their distances equal or unequal? Did any tie bind them to the earth, or to the sun, or to each other? Were their distances from the earth constant or variable? Were their motions irregular, or guided by law? Did they accomplish their revolutions among the fixed stars in regular curves, or in lawless wanderings? Among all the moving bodies, sun, moon, and planets, could any principle of association be traced which might bind them together and form them into a common system?

To resolve these profound questions, a critical watch is kept on all the moving bodies. Their pathway is among the stars, and to these ever during points of light their positions are constantly referred. If beyond the limits of the moving bodies a dark veil had been drawn so as to have excluded the light of the stars, at the first glance it might seem that by such a change, simplicity would have been introduced, and the perplexity arising from the motion of the planets among the profusely scattered stars, would have been removed. But let us not judge too hastily. Blot out the stars, and give to the sun, moon and planets a blank heavens in which to move, and the possibility of unraveling their mysterious motions, mutual relations, and common laws, is gone forever.

This will become manifest when we reflect that on such a change, not a fixed point in all the heavens would remain, to which we could refer a moving planet. They must then be referred to each other, and the motion due to the one, would become inextricably involved in that due to the other, and neither could be determined with any precision. Like the ocean islands which guided the early mariners, so God has given to us the stars of heaven as the fixed points to which we can ever refer, in all parts of their revolutions. the places of the wandering planets, and the swiftly revolving moon.

As the necessity for accuracy in watching the movements of the planets became more apparent, the attention was directed to the acquisition of the means by which this might be accomplished. Hence we find in the earliest ages the astronomer grouping the fixed stars into constellations—breaking up the great sphere of the heavens into fragments, the more easily to study its parts in detail. Not only are the stars of each constellation numbered, their brilliancy noted, but their relative places in the constellation and to each other, are fixed with all the precision which the rude means then in use permitted. Names are fixed to these different groupings, when or where or by whom we know not. Neither history nor tradition lead us back to this first breaking up of the heavens, but the names then bestowed on the fragmentary parts, the richer constellations, have survived the fall of empires, and are fixed forever in the heavens.

Possessing now a thorough knowledge of the objects among which the planets were moving, and the means of measuring with approximate accuracy, their distance from the stars along their path, it became possible to trace a planet in its career, and to note the changes of its velocity. New and wonderful discoveries were thus made. It was found that all the planets moved with an irregular velocity. Sometimes swiftly advancing among the fixed stars, then slowly relaxing their speed, they actually stopped, turned backward in their career, stopped again, and then, at first slowly but afterwards more rapidly, resumed their onward motion. These strange and anomalous motions, differing from anything remarked in the sun and moon, furnished new themes for discussion, new problems for solution. While the phenomena above alluded to became known, the same chain of observations revealed the remarkable fact, that the periods of revolution of the planets, though differing for each one of the group, were identical for any one individual, and moreover, that a simple curve marked out the pathway of sun, moon and planets, among the fixed stars, and that all these wandering bodies were confined to a narrow zone or belt in the heavens.

Centuries had now rolled away, nay, even thousands of years had slowly glided by, since the mind had first given itself to the examination of the heavens, and while discovery after discovery had rewarded the zeal of the observer in every age, yet the grand object of research, the distinction between actual and apparent motion, had thus far eluded the utmost efforts of human genius. But a brighter day was dawning. Each successive effort tore away some petty obstruction which impeded the march of mind upward towards the lofty region of truth. Facts grew and multiplied. Phenomena striking and diversified, were collated and compared. The mind in imagination took leave of the earth as the centre of all these complex movements, inexplicable on its surface, and naturally urged its flight towards the sun. There it paused and rested, and from this fixed point looked out upon the circling orbs, and lo! the complexity of their movements melted away. The centre was found—the mystery solved—the ponderous earth rescued from its false position, rolled in its place among the planets, one of the great family that swept in beauty and harmony about their common parent the sun.

The mind now stood upon the first platform of the rocky pyramid which it had been slowly rearing and with which it had been slowly rising, through long centuries of ceaseless toil. One grand point had been gained. Darkness had given way to light, but the great problem of the universe was yet to be resolved. All this long and arduous struggle had only revealed what the problem was. Appearances were now separated from realities, and with a fresh and invigorated courage the human mind now gives its energies to the accomplishment of definite objects, no longer working uncertainly in the dark, but with the clear light of truth to guide and conduct the investigation.

Possessed of these extraordinary advantages, the advance now became rapid and brilliant, as it had previously been slow and discouraging. That the planets, reckoning the earth as one, constituted a mighty family of worlds, was now manifest—whether linked singly to the sun, or mutually influencing each other, was the grand question. This great problem rested upon the resolution of a multitude of subordinate ones. The actual curve constituting the planetary orbits, the magnitude of these orbits, their actual position in space, the values and directions of their principal lines, the laws of their motion, all these and many more questions of equal importance and intricacy presented themselves in the outset of the examination now fairly commenced. Human skill was exhausted in the contrivance and construction of mechanical aids by which the movements of the planets might be watched with the greater accuracy. Partial success crowned these extraordinary efforts, but there yet remained delicate investigations which with the utmost skill in observing escaped the farthest reach of man's eagle gaze, and seemed to bid defiance to all his powers.

To conquer these difficulties, one of two things must be accomplished: either man must sweep out from earth towards the distant planets, to gain a nearer and more accurate view, or else bring them down from their lofty spheres to subject themselves to his scrutinizing gaze. How hopeless the accomplishment of either of these impossible alternatives. But who shall prescribe the limits of human genius? In studying the phenomena of the passage of light through transparent crystalized bodies, a principle is discovered which lets in a gleam of hope on the disheartened mind. It seizes this principle, converts it to its use, and arms itself with an instrument more wonderful than any that fancy in its wildest dreams ever pictured to the imagination. With the potent aid of this magic instrument, the astronomer no longer is bound hopelessly to his native earth; without indeed quitting in person its surface, his eye gifted with superhuman power, ranges the illimitable fields of space. He visits the moon, and finds a world with its lofty mountains and spreading valleys. The star-like planets swell into central worlds, with their circling moons, and myriads of fixed stars, hitherto beyond the reach of human vision, stand revealed in all their sparkling beauty. It is as if the united ranges of a thousand eyes were all concentrated in a single one.

A new era now dawns on the world. The delicate and invisible irregularities of the planetary motions are now fully revealed, and the data rapidly accumulate by means of which the last grand question is to be resolved. The orbital curves are determined. The laws of the revolving planets are revealed. A mysterious relation between the distances of the planets from the sun and their periods of revolution unites them positively into one grand family group. That they are bound to the sun by some inscrutable power, is certain, and it now remains to determine the law of increase and decrease of this force for all possible distances. This last truth is finally achieved, and the wisdom of God is vindicated in the beautiful structure of our grand system.

The second lofty platform is reached in the mighty pyramid, whose summit is now nearing the stars of heaven. From this elevation the mind looks out upon the circling planets and their revolving satellites, and the mysterious comet, and ventures to propound the question, do these bodies so interfere with the movements of each other, as to effect permanently the structure by which the equilibrium and stability of the entire system is guarantied?

To answer this question, a new train of investigation is commenced, satellite is weighed against planet, and planet against the sun, until the mass of matter contained in each individual of the system becomes accurately known. Then is undertaken the grand problem of perturbations. The telescope reveals the fact that slow and mysterious changes are going on in the mean motions of the moon, in the figure of the planetary orbits, and in the relative positions which these orbits hold to each other. Are these change ever progressive? If this be true, then, does the system contain within itself the seeds of decay, the elements of its own destruction. Slowly but surely as the solemn tread of time, the end must come, and one by one planet and satellite and comet, sink forever in the sun. Long and arduous was the struggle to reach the true answer to this difficult question. The entire solution involved a multitude of parts.

When the mutual dependence of the multitude of bodies constituting our system was discovered, when planet, and satellite, and comet, were found to feel and sway to the influence which each exerted on the other, the simplicity of their movements was gone forever; orbits once fixed in the heavens, slowly swung away from their moorings; the beautiful precision which had to all appearance marked the planetary curves, was destroyed. The regularity of their motions was changed into irregularity and a system of complexity which seemed to bid defiance to all effort at comprehension, presented itself to the human intellect.

It was no less than this—given, a system of revolving worlds, mutually operating on each other; required, their magnitude, masses, distances, motions, and positions, at the close of a thousand revolutions. What mind possessed the gigantic power to grasp this mighty problem? Reason was lost in wandering mazes, and the brightest intellect sunk clouded in gloom.

In this dilemma, the mind turns inward on its own resources. As the physical man climbs some mountain height by successive efforts, rising higher and higher, scaling rock after rock, and mounting precipice after precipice, by the use of strength comparatively feeble, resting and recruiting as it becomes exhausted, was it impossible, to contrive some mental machinery which might give to the reason the power of prosecuting its difficult researches, in such manner that it might stop and rest and not lose what it had already gained in its onward movement?

Geometry had invigorated the reason, as exercise toughens and strengthens the muscles of the human frame. But it had given to the mind no mechanical power, wherewith to conquer the difficulties which rose superior to its natural strength. Archimedes wanted but a place whereon to stand, and with his potent lever he would lift the world. The astronomer demands an analogous mental machinery to trace out the complex wanderings of a system of worlds. What the human mind demands and resolves to find, it never fails to discover. The infinitesimal analysis is reached, its principles developed, its resistless power, compelled into the service of human reason. I shall not now stop to explain the nature of this analysis. Its power and capacity alone engage our attention at the present. Once having seized on a wandering planet, it never relaxes its hold, no matter how complicated its movements, how various the influences to which it may be subjected, how numerous its revolutions, no escape is possible. This subtle analysis clings to its object, tracing its path and fixing its place with equal ease, at the beginning, middle, or close of a thousand revolutions, though each of these should require a century for its accomplishment.

Armed with this analysis, which the mind had created for its use, giving to it a strength only commensurate with the increased power which had been given to the human eyes, it concentrated its energies once more upon this last greatest problem. One by one these strongholds give way, the resistless power of analysis marches onward from victory to victory, until finally the sublime result is reached, the system is stable, the equilibrium is perfect; slowly rocking to and fro in periods which stun the imagination, the limits are prescribed beyond which these fluctuations shall never pass.

Here it would seem that human ambition might rest. Satisfied with having mastered the mysteries of the system with which we are united, the mind might cease its arduous struggle, and leave the wilderness of fixed stars free from its intrusions and ceaseless persecutions. But this is not the effect produced by victory; success but engenders new desires, and prompts to more difficult enterprises. Man having obtained the mastery over his own system, boldly wings his flight to the star-lit vault, and resolves to number its countless millions, to circumscribe its limitless extent, to fathom its infinite depth, to fix the centre about which his innumerable host is wheeling its silent and mysterious round.

Here commences a new era. The first step in the stupendous enterprise is to determine the distance of some one fixed star. Here again the mind is long left to struggle with difficulties which it seemed that no ingenuity or skill could remove. But its efforts do not go unrewarded. If it fails in the accomplishment of its grand object, it is rewarded by the most brilliant discoveries. The mighty law governing the planetary worlds is extended to the region of the fixed stars, motion is there detected, orbital motion, the revolution of sun about sun. The swift velocity of light is measured, to become the future unit in the expression of the mighty distances which remain yet to be revealed. Ever baffled but never conquered, the mind returns again and again to the attack, till finally the problem slowly yields, the immeasurable gulf is passed, and the distance of a single star rewards the toils of half a century. But what a triumph is this? It is no less than a revelation of the scale on which the universe is built. The interval from sun to fixed stars, is that by which the stars are separated ; and a reach of distance is opened up to the mind, which it only learns to contemplate by long continued effort.

But another startling fact is revealed in the prosecution of these profound investigations. The minute examinations of the fixed stars, have changed their character. For thousands of years they had been regarded as absolutely fixed among each other. This proves to be mere illusion, resulting from the use of means inadequate for the determination of their minute changes. Under the scrutinizing gaze of the eye, with its power increased a thousand-fold, the millions of shining orbs which fill the heavens, are all found to be slowly moving around each other, slowly as seen from our remote position, but with amazing velocity when examined near at hand.

A new problem of surprising grandeur now presents itself. Are these motions real? or are they due to a motion in the great centre of our system? A series of examinations analogous to those which divided between the real and apparent motions in the planets, is commenced and prosecuted with a zeal and devotion unsurpassed in the history of science. The mind rises to meet the sublime investigation. For a hundred years it toils on; again it triumphs; the truth is revealed. The immobility of the sun is gone forever; our last fixed point is swept from under us, and now the entire universe is in motion.

With redoubled energy the mind still prosecutes the inquiry, whither is the sun sweeping, and with what velocity does it pursue its unknown path? Strange and incredible as it may appear, these questions are answered; and in attaining this answer, the means are reached to separate between the real and apparent motions of the fixed stars, and to study their complex changes, and to rise by slow degrees, to a complete knowledge of the movements of the grand sidereal system. Here we pause. Rapidly have we descended the current of astronomical research, we have attained the boundary of the known. We stand on the dim confines of the unknown. All behind us is clear and bright and perfect, all before us is shrouded in gloom and darkness and doubt. Yet the twilight of the known flings its feeble light into the domain of the unknown, and we are permitted to gather some idea, not of all that remains to be done, but of that which must be first accomplished.

Let us then stretch forward and propound some of those questions which nature yet presents for solution, but which have hitherto resisted the efforts of the human mind. First of all, we begin with our own system. How came it to be constituted as it actually exists? All the analogies of nature forbid the idea that it was thus instantly called into being by the fiat of Omnipotence. Does it come, then, from some primitive modification of matter, under the action of laws working out their results in countless millions of ages? Who shall present the true cosmogony of the solar system.

But this is only one unit among many millions. Whence the myriads of stars? those stupendous aggregations into mighty clusters? what the laws of their wonderful movements, of their perpetual stability? Who will explain the periodical stars, that wax and wane, like the changing moon: or still more wonderful, reveal the mystery of those which have suddenly burst on the astonished vision of man, and have as suddenly gone out forever in utter darkness.

Such are the questions which remain for the resolution of future ages. We may not live to witness these anticipated triumphs of mind over matter; but who can doubt the final result? Look backward to the Chaldean shepherd, who watched the changing moon from the plains of Shinar, and wondering, asked if future generations would reveal those mysterious phases? Compare his mind and knowledge with those of the modern astronomer, who grasps at a single glance, the past, present, and future changes of an entire system. Are the heights which remain to be reached, more rugged, more inaccessible, than those which have been already so triumphantly scaled? The observations recorded in Babylon three thousand years ago, have reached down through the long series of centuries, and are of inestimable value, in the solution of some of the darkest problems with which the mind has ever grappled. In like manner, the records we are now making, shall descend to unborn generations, and contribute to effect the triumphs of genius when three thousand years shall have rolled away. If doubt arises as to the final resolution of these profound questions, from the immense distance of the objects under examination, let us call to mind the fact, that the artificial eye which man has furnished for his use, possesses a glance so piercing, that no distance can hide an object from his searching vision.

Should Sirius, to escape this fiery glance, dart away from its sphere, and wing its flight at a velocity of twelve millions of miles in every minute, for a thousand years; nay, should it sweep onward at the same speed for ten thousand years, this stupendous distance cannot bury it from the persecuting gaze of man. But if distance is to form no barrier, no terminus to these investigations, surely there is one element which no human ingenuity can overcome. The complex movements of the planetary orbs have been revealed, because they have been repeated a thousand times under the eye of man, and from a comparison of many revolutions, the truth has been evolved. But tens of thousands of years must roll away before the most swiftly moving of all the fixed stars shall complete even a small fragment of its mighty orbit. With motions thus shrouded, these would seem to be in entire security from the inquisitive research of a being whose whole sweep of existence is but a moment, when compared with these vast periods. But let us not judge too hastily. The same piercing vision that follows the retreating star to depths of space almost infinite, is armed with a power so great, that if this same star should commence to revolve around some grand centre, and move so slowly that five millions of years must roll away before it can complete one circuit, not even a single year shall pass before its motion shall be detected, in ten years its velocity shall be revealed, and in the life-time of a single observer its mighty period shall become known.

If human genius is not to be baffled either by distance or time, numbers shall overwhelm it, and the stars shall find their safety in their innumerable millions. This retreat may even fail. The watch towers of science now cover the whole earth, and the sentinels never sleep. No star or cluster or constellation, can ever set. It escapes the scrutinizing gaze of one astronomer, to meet the equally piercing glance of another. East and west and north and south, from the watch towers of the four quarters of the globe, peals the solemn mandate, onward!

Here we pause. We have closed the enunciation of the great problem whose discussion and solution lie before us, a problem whose solution has been in progress six thousand years—one which has furnished to man the opportunities of his loftiest triumphs, one which has taxed in every age the most vigorous efforts of human genius, a problem whose successive developments have demonstrated the immortality of mind and whose sublime results have vindicated the wisdom and have declared the glory of God. You have listened to the enunciation, we now invite you to follow us in the demonstration. And may that Almighty power, which built the heavens, gave to me wisdom to reveal, and to you power to grasp the truths and doctrines, wrested by mind from nature in its long struggle of sixty centuries of toil!

LECTURE II.

THE DISCOVERIES OF THE PRIMITIVE AGES.

To those who have given but little attention to the science of astronomy, its truths, its predictions, its revelations, are astonishing; and but for their rigorous verification, would be absolutely incredible. When we look out upon the multitude of stars which adorn the nocturnal heavens, scattered in bright profusion in all directions, without law, and regardless of order—when with telescopic aid, thousands are increased to millions, and suns and systems and universes, rise in sublime perspective, as the visual ray sweeps outwards to distances which defy the powers of arithmetic to express, how utterly futile does it seem, for the mind to dare to pierce and penetrate, to number, weigh, measure and circumscribe, these innumerable millions? It is only when we remember, that from the very cradle of our race, strong and powerful minds, have in rapid and continuous succession, bent their energies upon the solution of this grand problem, that we can comprehend, how it is, that light now breaks in upon us, from the very confines of the universe, dimly revealing the mysterious forms, which lie yet half concealed in the unfathomable gulfs of space. When I reflect on the recent truimphs of human genius—when I stand on the shore of that mighty stream of discovery, which has grown broader and deeper, as successive centuries have rolled away, gathering in strength and intensity, until it has embraced the whole universe of God; I am carried backward through thousands of years, following this stream, as it contracts towards its source, till finally its silver thread is lost in the clouds and mists of antiquity. I would fain stand at the very source of discovery, and commune with that unknown god-like mind which first conceived the grand thought, that even these mysterious stars might be read, and that the bright page which was nightly unfolded to the vision of man, needed no interpreter of its solemn beauties, but human genius. There is to my mind, no finer specimen of moral grandeur, than that presented by him who first resolved to read and comprehend the heavens. On some lofty peak he stood, in the stillness of the midnight hour, with the listening stars as witnesses of his vows, and there, conscious of his high destiny, and of that of his race, resolves to commence the work of ages. "Here," he exclaims, "is my watch-tower, and yonder bright orbs, are henceforth my solitary companions. Night after night, year after year, will I watch and wait, ponder and reflect, until some ray shall pierce the deep gloom which now wraps the world."

Thus resolved the unknown founder of the science of the stars. His name and his country are lost forever. What matters this, since his works, his discoveries, have endured for thousands of years and will endure, as long as the moon shall continue to fill her silver horn, and the planets to roll and shine.

Go with me, then, in imagination, and let us stand beside this primitive observer, at the close of his career of nearly a thousand years, (for we must pass beyond the epoch of the deluge, and seek our first discoveries among those sages, whom, for their virtues, God permitted to count their age, not by years, but by centuries,) and here we shall learn the order in which the secrets of the starry world slowly yielded themselves, to long and persevering scrutiny. And now let me unfold, in plain and simple language, the train of thought, of reasoning and research, which marked this primitive era of astronomical science. It is true that history yields no light, and tradition even fails, but such is the beautiful order in the golden chain of discovery, that the bright links which are known, reveal with certainty, those which are buried in the voiceless past.—If then it were possible to read the records of the founder of astronomy, graven on some column of granite, dug from the earth, whither it had been borne by the fury of the deluge, we know now what its hieroglyphics would reveal, with a certainty scarcely less than that which would be given by an actual discovery, such as we have imagined. We are certain that the first discovery ever recorded, as the result of human observation, was on the *moon*.

The sun, the moon, the stars, had long continued to rise, and climb the heavens, and slowly sink beneath the western horizon. The spectacle of day and night, was then as now, familiar to every eye; but in gazing there was no observation, and in mute wonder there was no science. When the solitary observer took his post, it was to watch the moon. Her extraordinary phases had long fixed his attention. Whence came these changes? The sun was ever round and brilliant—the stars shone with undimmed splendor—while the moon was ever waxing and waning, sometimes a silver crescent hanging in the western sky, or full orbed, walking in majesty among the stars, and eclipsing their radiance, with her overwhelming splendor. Scarcely had the second observation been made upon the moon, when the observer was struck with the wonderful fact, that she had left her place among the fixed stars, which on the preceding night he had accurately marked. Astonished, he again fixes her place by certain bright stars close to her position, and waits the coming of the following night. His suspicions are confirmed—the moon is moving; and what to him is far more wonderful, her motion is precisely *contrary* to the general revolution of the heavens, from east to west. With a curiosity deeply aroused, he watches from night to night, to learn whether she will return upon her track; but she marches steadily onward among the stars, until she sweeps the entire circuit of the heavens, and returns to the **point first occupied, to recommence her ceaseless cycles.**

An inquiry now arose, whether the changes in the moon, her increase and decrease, could in any way depend on her place among the fixed stars. To solve this question required a longer period. The group of stars among which the new moon was first seen was accurately noted, so as to be recognized at the following new moon, and doubtless our primitive astronomer hoped to find that in this same group the silver crescent, when it should next appear, would be found. But in this he was disappointed; for when the moon became first faintly visible in the western sky, the group of stars which had ushered her in before, had disappeared below the horizon, and a new group had taken its place; and thus it was discovered that each successive new moon fell farther and farther backward among the stars. By counting the days from new moon to new moon, and those which elapsed while the moon was passing round the heavens from a certain fixed star to this same star again, it was found, that these two periods were different; the revolution from new to new occupying 29½ days, while the sidereal revolution, from star to star, required 27½ days.

This backward motion of the moon among the stars, must have perplexed the early astronomers; and for a long while it was utterly impossible to decide whether the motion was real or only apparent—analogy would lead to the conclusion that all motion must be in the same direction, and as the heavens revolved from east to west, it seemed impossible that the moon, which manifestly participated in this general movement, should have another and a different motion, from west to east. There was one solution of this mystery, and I have no doubt it was for a long while accepted and believed. It was this. By giving to the moon a slower motion from east to west, than the general motion of the heavens, she would appear to lag behind the stars, which would by their swifter velocity pass by her, and thus occasion in her the observed apparent motion, from west to east. We shall see presently how this error was detected.

The long and accurate vigils of the moon, and the necessity of recognizing her place, by the clusters of groups of stars among which she was nightly found, had already familiarized the eye with those along her track, and even thus early the heavens began to be divided into constellations. The eye was not long in detecting the singular fact, that this stream of constellations, lying along the moon's path, was constantly flowing to the west, and one group after another apparently dropping into the sun, or at least becoming invisible in consequence of their proximity to this brilliant orb. A closer examination revealed the fact, that the aspect of the whole heavens was changing from month to month. Constellations which had been conspicuous in the west, and whose brighter stars were the first to appear as the twilight faded, were found to sink lower and lower towards the horizon, till they were no longer seen; while new groups were constantly appearing in the east.

These wonderful changes, so strange and inexplicable, must have long perplexed the early student of the heavens. Hitherto the stars along the moon's route, had engaged special attention; but at length certain bright and conspicuous constellations, towards the north, arrested the eye: and these were watched to see whether they would disappear.—Some were found to dip below the western horizon, soon to re-appear in the east; while others revolving with the general heavens, rose high above the horizon, swept steadily round, sunk far down, but never disappeared from the sight. This remarkable discovery soon led to another equally important. In watching the stars in the north through an entire night, they all seemed to describe circles; having a common center, these circles grew smaller and smaller as the stars approached nearer to the center of revolution, until finally one bright star was found, whose position was ever fixed—Alone unchanged while all else was slowly moving. The discovery of this remarkable star, must have been hailed with uncommon delight by the primitive observer of the heavens. If his deep devotion to the study of the skies, had created surprise among his rude countrymen, when he came to point them to this never changing light hung up in the heavens, and explained its uses in guiding their wanderings on the earth, their surprise must have given place to admiration. Here was the first valuable gift of primitive astronomical science to man.

But to the astronomer this discovery opened up a new field of investigation, and light began to dawn on some of the most mysterious questions which had long perplexed him. He had watched the constellations near the moon's track slowly disappear in the effulgence of the sun, and when they were next seen, it was in the east, in the early dawn, apparently emerging from the solar beams, having actually passed by the sun. Watching and reflecting, steadily pursuing the march of the northern constellations, which never entirely disappeared, and noting the relative positions of these, and those falling into the sun, it was at last discovered that the entire starry heavens was slowly moving forward to meet and pass by the sun, or else the sun itself was actually moving backward among the stars. This apparent motion had already been detected in the moon, and now came the reward of long and diligent perseverance. The grand discovery was made, that both the sun and moon were moving among the fixed stars, not *apparently*, but *absolutely*. The previously received explanation of the moon's motion, could no longer be sustained; for the starry heavens could not at the same time so move as to pass by the moon in one month, and to past by the sun in a period twelve times as great. A train of the most important conclusions flowed at once from this great discovery.—The starry heavens passed beneath and around the earth,—the sun and moon were wandering in the same direction, but with different velocities among the stars,—the constellations actually filled the entire heavens above the earth and beneath the earth,—the stars were invisible in the day time, not because they did not exist, but because their feeble light was lost in the superior brilliancy of the sun. The heavens were spherical, and encompassed like a shell the entire earth, and hence it was conceived that the earth itself was also a globe, occupying the center of the starry sphere.

It is impossible for us, familiar as we are at this day with these important truths, to appreciate the rare merit of him who by the power of his genius, first rose to their knowledge and revealed them to an astoished world. We delight to honor the names of Kepler, of Galileo, of Newton; but here are discoveries so far back in the dim past, that all trace of their origin is lost, which vie in interest and importance with the proudest achievements of any age.

With a knowledge of the sphericity of the heavens, the revolution of the sun and moon, the constellations of the celestial sphere, the axis of its diurnal revolution, astronomy began to be a science, and its future progress was destined to be rapid and brilliant. A line drawn from the earth's center to the north star formed the axis of the heavens, and day and night around this axis all the celestial host were noiselessly pursuing their never ending journeys.—Thus far, the only moving bodies known, were the sun and moon. These large and brilliant bodies, by their magnitude and splendor, stood out conspicuously, from among the multitude of stars, leaving these minute but beautiful points of light, in one great class, unchangeable among themselves, fixed in their grouping and configurations, furnishing admirable points of reference, in watching and tracing out the wanderings of the sun and moon.

To follow the moon as she pursued her journey among the stars was not difficult; but to trace the sun in his slower and more majestic motion, and to mark accurately his track, from star to star, as he heaved upward to meet the coming constellations, was not so readily accomplished. Night after night, as he sunk below the horizon, the attentive watcher marked the bright stars near the point of setting which first appeared in the evening twilight.—These gradually sunk towards the sun on successive nights, and thus was he traced from constellation to constellation, until the entire circuit of the heavens was performed, and he was once more attended by the same bright stars, that had watched long before, his sinking in the west. Here was revealed the measure of the *Year*. The earth had been verdant with the beauties of spring,—glowing with the maturity of summer,—rich in the fruits of autumn,—and locked in the icy chains of winter, while the sun had circled round the heavens. His entrance into certain constellations marked the coming seasons, and man was beginning to couple his cycle of pursuits on earth with the revolutions of the celestial orbs.

While intently engaged in watching the sun as it slowly heaved up to meet the constellations, some ardent devotee to this infant science, at length marked in the early twilight a certain brilliant star closely attendant upon the sun. The relative position of these two objects was noted, for a few consecutive nights, when with a degree of astonishment of which we can form no conception, he discovered that this brilliant star was rapidly approaching the sun, and actually changing its place among the neighboring stars,—night after night he gazes on this unprecedented phenomenon, *a moving star!* and on each successive night he finds the wanderer coming nearer and nearer to the sun. At last it disappears from sight, plunged in the beams of the upheaving sun. What had become of this strange wanderer? was it lost forever? were questions which were easier asked than answered.—But patient watching had revealed the fact, that when a group of stars, absorbed into the sun's rays, disappeared in the west, they were next seen in the eastern sky, slowly emerging from his morning beams. Might it not be possible, that this wandering star would pass by the sun and re-appear in the east? With how much anxiety must this primitive discoverer have watched in the morning twilight? Day after day he sought his solitary post, and marked the rising stars, slowly lifting themselves above the eastern horizon. The gray dawn came, and the sun shot forth a flood of light, the stars faded and disappeared, and the watcher gives over, till the coming morning. But his hopes are crowned at last. Just before the sun breaks above the horizon, in the rosy east, refulgent with the coming day, he descries the pure white silver ray of his long lost wanderer. It has passed the sun,—it rises in the east,—the first *planet* is discovered!

With how much anxiety and interest did the delighted discoverer trace the movements of his wandering star. Here was a new theme for thought, for observation, for investigation; would this first planet sweep round the heavens, as did the sun and moon? would it always move in the same direction? would its path lie among those groups of stars among which the sun and moon held their course? Encouraged by past success, he rejoicingly enters on the investigations of these questions. For some time the planet pursues its journey from the sun, leaving it farther and farther behind. But directly it slackens its pace,—it actually stops in its career, and the astonished observer perhaps thinks that his wandering star had again become fixed. Not so,—a few days of watching dispels this idea. Slowly at first and soon more swiftly, the planet seeks again the sun, moving backwards on its former path, until finally its light is but just visible in the east at early dawn. Again it is lost in the sun's beams for a time, and contrary to all preceding analogy, when next seen, its silver ray comes out pure and bright, just above the setting sun. It now recedes from the sun, on each successive evening increasing its distance, till it again reaches a point never to be passed—here it stops—is stationary for a day or two, and then again sinks downward to meet the sun. How wonderful and inexplicable the movements of this wandering star must have appeared in the

THE DISCOVERIES OF THE PRIMITIVE AGES. 31

early ages! oscillating backward and forward, never passing its prescribed limits, and ever closely attendant upon the sun. Where the sun sunk to repose, there did the faithful planet sink, and where the sun rose, at the same point did the wandering star make its appearance. The number of days was accurately noted, from the stationary point in the east above the sun, to the stationary point in the west above the sun, and thus the period, 584 days from station to station, became known.

The discovery of one planet, led the way to the rapid discovery of several others. If we may judge of their order by their brilliancy, Jupiter was the second wanderer revealed among the stars. Then followed Mars, and Saturn, and after a long interval Mercury was detected, hovering near the sun, and imitating the curious motions of Venus.

Here the progress of planetary discovery was suddenly arrested, keen as was the vision of the old astronomer, long and patient as was his scrutiny, no depth of penetration of unaided vision could stretch beyond the mighty orbit of Saturn, and the search was given over.—A close examination of the planets revealed many important facts. Three of them, Mars, Jupiter, and Saturn, were found to perform the circuit of the heavens, like the sun and moon, and in the same direction; with this remarkable difference, that while the sun and moon, moved steadily and uniformly in the same direction, the planets occasionally slackened their pace, would then stop, move backwards on their track, stop again, and finally resume their onward motion. Their periods of revolution were discovered by marking the time which elapsed, after setting out from some brilliant and well known fixed star, until they should perform the entire circuit of the heavens and once more return to the same star. The times of revolution were found to differ widely from each other; Mars requiring about 687 days, Jupiter 4,332 days, and Saturn 10,759 days, or nearly thirty of our years.

The planets all pursued their journeys in the heavens, among the same constellations which marked the paths of the sun and moon, and hence these groups of stars concentrated the greatest amount of attention among the early astronomers, and became distinguished from all the others.

Whatever light may be shed upon antiquity by deciphering the hieroglyphic memorials of the past there is no hope of ever going far enough back, to reach even the nation, to which we are indebted for the first rudiments of the science of the stars.

Thus far in the prosecution of the study of the heavens, the eye and the intellect had accomplished the entire work. Rapidly as we have sketched the progress of early discovery, and short as may have been the period in which it was accomplished, no one can fail to perceive, how vast is the difference between the light that thus early broke in upon the mind, heralding the coming of a brighter day, and the deep and universal darkness which had covered the world, before the dawn of science. Encouraged by the success which had thus far rewarded patient toil, the mind of man pushes on its investigations deeper and deeper into the domain of the mysterious and unknown.

In watching the annual revolution of the sun among the fixed stars, one remarkable peculiarity had long been recognized. While the interval of time, from the rising to the setting of the stars, was ever the same at all seasons of the year, the interval from the rising to the setting of the sun was perpetually changing, passing through a cycle which required exactly one year for its completion. It became manifest that the sun did not prosecute its annual journey among the stars, in a circle parallel with those described by the stars, in their diurnal revolution. His path was oblique to those circles, and while he participated in their diurnal motion, he was sweeping by his annual revolution round the heavens, and was at the same time, by another most extraordinary movement, carried towards the north to a certain distance, then stopping, commenced a return towards the south,—reached his southern limit,—again changed his direction, and thus oscillated from one side to the other of his mean position.

These wonderful changes became the objects of earnest investigation. In what curve did the sun travel among the stars? All diurnal motion was performed in a circle, the first discovered, the simplest and the most beautiful of curves: and in this curve, analogy taught the early astronomers, that all celestial movements must be performed. It became therefore, a matter of deep interest, to trace the sun's path accurately among the stars, to mark his track, and to see whether it would not prove to be a circle. To accomplish this, more accurate means must be adopted, than the mere watching of the stars which attended the rising or setting sun. The increase and decrease of the shadow of some high pointed rock, to whose refreshing shade the shepherd astronomer had repaired in the heat of noon, and beneath which he had long pondered this important problem, first suggested the means of its resolution.—As the summer came on he remarked that the length of the noon shadow of his rock, perpetually decreased from day to day. As the sun became more nearly vertical at noon, the shadow gave him less and less shelter. Watching these noon shadows, from day to day, he found them proportioned to the sun's northern or southern motion, and finally the thought entered his mind, that these shadows would mark with certainty the limits of the sun's motion north and south,—the character of his orbit or route among the stars,—the changes and duration of the seasons, and the actual length of the year, which thus far had been but roughly determined. To accomplish the observations more accurately, an area on the ground was smoothed and leveled, and in its centre a vertical pole was erected some ten or fifteen feet in length, whose sharp vertex cast a well defined shadow. And here we have the first astronomical instru-

THE DISCOVERIES OF THE PRIMITIVE AGES.

ment, the gnomon, ever devised by the ingenuity of man. Simple as it is, by its aid the most valuable results were obtained.

The great point was to mark with accuracy the length of the noon-day shadow, from month to month, throughout the entire year. Four remarkable points in the sun's annual track, were very soon detected and marked. One of these occurred in the summer, and was that point occupied by the sun on the day of the shortest noon shadow. Here the sun had reached his greatest northern point, and for a few days the noon shadow, cast by the gnomon appeared to remain the same, and the sun *stood still.* The noon shadows now increased slowly, for six months, as the sun moved south, till a second point was noted, when the noon shadow had reached its greatest length. Again it became stationary, and again the sun paused and *stood still,* before commencing his return towards the north. These points were called the *summer* and *winter solstices,* and occurred at intervals of half a year. At the summer solstice, the longest day occurred, while at the winter solstice, the shortest day was always observed. These extreme differences between the length of the day and night, occasioned the determination of the other two points. From the winter solstice, the noon shadows decreased as the length of the day increased, until finally the day and night were remarked to be of *equal* length, and the distance to which the shadow of the gnomon was thrown on that day, was accurately fixed. If on this day the diurnal circle described by the sun, could have been marked in the heavens by a circle of light, sweeping from the east to the west, so that the eye might rest upon and retain it, and if at the same time the sun's annual path, among the fixed stars could have been equally exhibited in the heavens by a circle of light, these two circles would have been seen to cross each other, and at their point of crossing, the sun would have been found. The diurnal circle was called the *equator,* the sun's path the *ecliptic,* and the point of intersection was called appropriately, the *equinox.*—As the sun crossed the equator in the spring and autumn, these points received the names of the *vernal* and *autumnal equinoxes* , and were marked with all the precision which the rude means then in use rendered practicable.

The bright circle already imagined in the heavens to represent the sun's annual track among the stars, passed obliquely across the equator, and the amount by which these circles were inclined to each other was actually measured in these early ages, with no mean precision by the noon shadows of the gnomon. The ray casting the shortest noon shadow, was inclined to the ray forming the longest noon shadows, under an angle precisely double of the inclination of the ecliptic or sun's path to the equator, and the inclination of these two rays marked exactly the annual motion of the sun from south to north, or from north to south. A close examination of the order of increase and decrease in the length of the noon shadows cast by the gnomon, demonstrated the important truth already suspected, that the sun's path was actually a *circle*, but inclined, as has already been shown, to the diurnal circles of the stars and to the equator.

By counting the days which elapsed from the summer solstice to the summer solstice again, a knowledge of the length of the year, or period of the sun's revolution, was obtained. But here again a discovery was made which produced an embarrassment to the early astronomer, which all their perseverance and research never succeeded in removing,

In these primitive ages the heavenly bodies were regarded with feelings little less than the reverence we now bestow on the Supreme Creator. The sun especially, as the Lord of life and light, was regarded with feelings nearly approaching to adoration, even by the astronomers themselves. The idea early became fixed, that the chief of the celestial bodies must move with a uniform velocity in a circular orbit, never increasing or decreasing. Change being inconsistent with the supreme and dignified station which was assigned to him—what then must have been the astonishment of the primitive astronomers, who in counting the days from the summer to the winter solstice, and from the winter round to the summer solstice, these intervals were found to be unequal.—This almost incredible result was confirmed, by remarking that the shortest spaces from equinox to solstice, dividing the sun's annual route, into four equal portions, were passed over in unequal times. These results could not be doubted, for each observation, from year to year, confirmed them. They were received and recorded, but the problem was handed down to succeeding generations for solution.

In consequence of the oblique direction of the ecliptic or sun's track, it was found difficult to retain its position in the mind. To assist in the recurrence to this important circle, a brazen circle was at length devised, and fastened permanently to another brazen circle of equal size, under an angle, exactly equal to the inclination of the equator to the ecliptic. Circles, perpendicular to the equator, and passing through the solstices and equinoxes, completed the second astronomical instrument, *the sphere*. Having constructed this simple piece of machinery, it was mounted on an axis passing through its centre, and perpendicular to its equator, so as to revolve, as did the heavens, whose motions it was intended to represent. Having so placed the axis of rotation, that its prolongation would pass through the north star, this rude sphere came to play a most important part in the future investigations of the heavens. Its brazen equator and ecliptic were each divided into a certain number of equal parts, by reference to which the motion of the heavenly bodies might be followed, with far greater precision, than had ever been previously obtained.

Armed with a new and more perfect instrument, the astronomer resumes his great investigation. Finding it now possible, to mark out the sun's path in the heavens, with certainty, by means of his brazen ecliptic, he discovers that the moon and planets in each revolution pass across the sun's track, and spend nearly an equal amount of time on the north and south sides of the ecliptic. This discovery led to a more accurate determination of the periods of revolution of the planets. The interval was noted from one passage across the ecliptic to the next on the same side, and these intervals marked with accuracy the planetary periods. It now became possible to fix, with greater certainty, the relative positions of the sun and moon, and problems were once more resumed which had thus far baffled every effort of human genius. The phases of the moon, the very first point of investigation, had never yet yielded up its hidden cause, and those terrific phenomena, solar and lunar eclipses, which had long covered the earth with terror and dismay, were wrapped in mystery, and their explanation had resisted the sagacity of the most powerful and gifted intellects.

No one has ever witnessed the going out of the sun in dim eclipse, even now when its most minute phenomena are predicted with rigorous exactitude, without a feeling of involuntary dismay. What then must have been the effect upon the human mind in those ages of the world, when the cause was unknown and when these terrific exhibitions burst on earth's inhabitants unheralded and unannounced? Here then was an investigation, not prompted by curiosity alone, but involving the peace and security of man in all coming ages. We cannot doubt that the causes of the solar eclipse were first detected. It was observed, that no eclipse of the sun ever occurred, when the moon was visible. Even during a solar eclipse, when the sun's light had entirely faded away, and the stars and planets stole gently upon the sight in the sombre and unnatural twilight, the moon was sought for in vain; she was never to be seen. This fact excited curiosity and gave rise to a careful and critical examination of the place in which the moon should be found, immediately after a solar eclipse; and it was soon discovered that on the night following the day of eclipse, the moon was seen in her crescent shape very near to the sun and but a short distance from the sun's path. By remarking the moon's places, next before a solar eclipse and that immediately following, it was seen that at the time of the occurrence of the eclipse, the moon was actually passing from the west to the east side of the sun's place, and finally a little calculation showed that a coincidence of the sun and moon in the heavens took place at the precise time at which the sun had been eclipsed. The conclusion was irresistible, and the great fact was announced to the world, that the sun's light *was hidden by the interposition of the dark body of the moon.*

Having reached this important result with entire certainty, the explanation of the moon's phases followed in rapid succession. For it now became manifest, that the moon shone with *borrowed light*, and that her brilliancy came from the reflected beams of the sun. This was readily demonstrated by the following facts. When the moon was so situated that the side next to the sun, (the illuminated one), was turned from the eye of the observer, (as was the case in a solar eclipse), then the moon's surface next to the observer, was always found to be entirely black. Pursuing her journey from this critical point, the moon was next seen near the sun, in the evening twilight, as a slender thread of light, a very small portion of her illuminated surface being now visible. Day after day this visible portion increases, until finally the moon rises as the sun sets, full orbed and round, being directly opposite the sun, and turning her entire illuminated surface towards the eye of the observer. By like degrees she loses her light as she approaches, and finally becomes invisible as she passes by the sun. From this examination it became evident that the moon was a globular body, non-luminous, and revolving in an orbit, comprehended entirely within that described by the sun, and consequently, nearer to the earth than the sun. Having ascertained this fact, it was concluded that among all the moving heavenly bodies, the periods of revolution indicated their relative distances from the earth. Hence Mars was regarded as more distant than the sun, Jupiter more remote than Mars, and Saturn the most distant, as it was the slowest moving of all the planets.

After reaching to a knowledge of the causes producing the eclipse of the sun and the phases of the moon, it remained yet to resolve the mystery of the lunar eclipse. It was far more difficult to render a satisfactory account of this phenomenon than either of the preceding. The light of the moon was not intercepted by the interposition of any opaque body, between it and the eye of the observer. No such body existed, and long and perplexing was the effort to explain this wonderful phenomenon. Finally it was observed that all opaque bodies cast shadows, in directions opposite to the source of light. Was it not possible that the light of the sun, falling upon the earth, might be intercepted by the earth, and thus produce a shadow which might even reach as far as the moon? So soon as this conjecture was made, a series of examinations were commenced to confirm or destroy the theory. It was at once seen that in case the conjecture was true, no lunar eclipse could occur except when the sun, earth and moon, were situated in the same straight line; a position which could never occur, except at the *full* or *new* of the moon. It was soon discovered that, it was only at the full, that lunar eclipses took place, thus confirming the truth of the theory, and fixing it beyond a doubt that the shadow of the earth falling on the moon, was the cause of her eclipse. The moon had already been shown to be non-luminous, and the moment the interposition of the earth, between it and its source of light, the sun, cut off its light, it ceased to be visible, and passed through an eclipse. The sphericity of the earth, which had been analogically inferred from that of the heavens was now made absolutely certain—for it was remarked, that as the moon entered the earth's shadow, that the track of this dark shadow across the bright surface of the moon was always *circular*, which was quite impossible, for every position except the earth, which cast this circular shadow, should be of a globular form.

Having now attained to a clear and satisfactory explanation of the two grand phenomena, solar and lunar eclipses, the question naturally arose, why was not the sun eclipsed in each revolution of the moon; and how happened it that the moon in the full, did not always pass through the earth's shadow? An examination of the moon's path among the fixed stars gave to these questions a clear and positive answer. It was found that the sun and moon did not perform their revolutions in the same plane. The moon's route among the stars crossed the sun's route under a certain angle, and it thus frequently happened, that at the new and full, the moon occupied some portion of her orbit too remote from that of the sun to render either a lunar or solar eclipse possible.

Rapidly have we traced the career of discovery. The toil and watching of centuries have been condensed into a few moments of time, and questions requiring ages for their solution have been asked, only to be be answered. In connection with the investigations just developed, and as a consequence of their successful prosecution, the query arose whether in case science had reached to a true exposition of the causes producing the eclipse of the sun, was it not possible to stretch forward in time, and anticipate and predict the coming of these dread phenomena?

To those who have given but little attention to the subject, even in our own day, with all the aids of modern science, the prediction of an eclipse, seems sufficiently mysterious and unintelligible. How then it was possible, thousands of years ago, to accomplish the same great object, without any just views of the structure of the system, seems utterly incredible. Follow me, then, while I attempt to reveal the train of reasoning which led to the prediction of the first eclipse of the sun, the most daring prophecy ever made by human genius. Follow in imagination, this bold interrogator of the skies to his solitary mountain summit—withdrawn from the world—surrounded by his mysterious circles, there to watch and ponder through the long nights of many—many years. But hope cheers him on, and smooths his rugged pathway. Dark and deep as is the problem, he sternly grapples with it, and resolves never to give over till victory crowns his efforts.

He has already remarked, that the moon's track in the heavens crossed the sun's, and that this point of crossing was in some way intimately connected with the coming of the dread eclipse. He determines to watch and learn whether the point of crossing was fixed, or whether the moon in each successive revolution, crossed the sun's path at a different point. If the sun in its annual revolution could leave behind him a track of fire marking his journey among the stars, it is found that this same track was followed from year to year, and from century to century with undeviating precision. But it was soon discovered, that it was far different with the moon. In case she too could leave behind her a silver thread of light sweeping round the heavens, in completing one revolution, this thread would not join, but would wind around among the stars in each revolution, crossing the sun's fiery track at a point west of the previous crossing. These points of crossing were called the *moon's nodes*. At each revolution the node occurred further west, until after a cycle of about nineteen years, it had circulated in the same direction entirely round the ecliptic. Long and patiently did the astronomer watch and wait, each eclipse is duly observed, and its attendant circumstances are recorded, when, at last, the darkness begins to give way and a ray of light breaks in upon his mind. He finds that no eclipse of the sun ever occurs unless the *new moon is in the act of crossing the sun's track*. Here was a grand discovery.—He holds the key which he believes will unlock the dread mystery and now, with redoubled energy, he resolves to thrust it into the wards and drive back the bolts.

To predict an eclipse of the sun, he must sweep forward, from new moon to new moon, until he finds some new *moon* which should occur, while the moon was in the act of crossing from one side to the other of the sun's track.—This certainly was possible. He knew the exact period from new moon to new moon, and from one crossing of the ecliptic to another. With eager eye he seizes the moon's place in the heavens, and her age, and rapidly computes where she will be at her next change. He finds the new moon occurring far from the sun's track; he runs round another revolution; the place of the new moon falls closer to the sun's path, and the next yet closer, until reaching forward with piercing intellectual vigor, he at last, finds a new moon which occurs precisely at the computed time of her passage across the sun's track. Here he makes his stand, and on the day of the occurrence of that new moon, he announces to the startled inhabitants of the world, that the sun shall expire in dark eclipse—Bold prediction !—Mysterious prophet ! with what scorn must the unthinking world have received this solemn declaration. How slowly do the moons roll away, and with what intense anxiety does the stern philosopher await the coming of that day which should crown him with victory, or dash him to the ground in ruin and disgrace. Time to him moves on leaden wings; day after day, and at last hour after hour, roll heavily away. The last night is gone—the moon has disappeared from his eagle gaze in her approach to the sun, and the dawn of the eventful day breaks in beauty on a slumbering world.

This daring man, stern in his faith, climbs alone to his rocky home, and greets the sun as he rises and mounts the heavens, scattering brightness and glory in his path. Beneath him is spread out the populous city, **already teeming with life and activity.** The busy morning hum rises on

THE DISCOVERIES OF THE PRIMITIVE AGES.

the still air and reaches the watching place of the solitary astronomer. The thousands below him, unconscious of his intense anxiety, buoyant with life, joyously pursue their rounds of business, their cycles of amusement. The sun slowly climbs the heavens, round and bright and full orbed. The lone tenant of the mountain-top almost begins to waver in the sternness of his faith, as the morning hours roll away. But the time of his triumph, long delayed, at length begin to dawns; a pale and sickly hue creeps over the face of nature. The sun has reached his highest point, but his splendor is dimmed, his light is feeble. At last it comes! —Blackness is eating away his round disk,—onward with slow but steady pace, the dark veil moves, blacker than a thousand nights—the gloom deepens,—the ghastly hue of death covers the universe,—the last ray is gone, and horror reigns. A wail of terror fills the murky air,—the clangor of brazen trumpets resounds,—an agony of despair dashes the stricken millions to the ground, while that lone man, erect on his rocky summit, with arms outstretched to heaven, pours forth the grateful gushings of his heart to God, who had crowned his efforts with triumphant victory. Search the records of our race, and point me, if you can, to a scene more grand, more beautiful. It is to me the proudest victory that genius ever won. It was the conquering of nature, of ignorance, of superstition, of terror, all at a single blow, and that blow struck by a single arm.—And now do you demand the name of this wonderful man! Alas! what a lesson of the instability of earthly fame are we taught in this simple recital.—He who had raised himself immeasurably above his race,—who must have been regarded by his fellows as little less than a god, who had inscribed his fame on the very heavens, and had written it in the sun, with a " pen of iron, and the point of a diamond," even this one has perished from the earth—name, age, country, are all swept into oblivion, but his proud achievement stands. The monument reared to his honor stands, and although the touch of time has effaced the lettering of his name, it is powerless, and cannot destroy the fruits of his victory.

A thousand years roll by: the astronomer stands on the watch tower of old Babylon, and writes for posterity the records of an eclipse; this record escapes destruction, and is safely wafted down the stream of time. A thousand years roll away: the old astronomer, surrounded by the fierce, but wondering Arab again writes, and marks the day which witnesses the sun's decay. A thousand years rolls heavily away: once more the astronomer writes from amidst the gay throng that crowds the brightest capital of Europe. Record is compared with record, date with date, revolution with revolution, the past and present are linked together,—another struggle commences, and another victory is won. Little did the Babylonian dream that he was observing for one who after the lapse of 3000 years, should rest upon records, the successful resolution of one of nature's darkest mysteries.

We have now reached the boundary where the stream of discovery, which we have been tracing through the clouds and mists of antiquity, begins to emerge into the twilight of tradition, soon to flow on in the clear light of a history that shall never die. Henceforth our task will be more pleasing, because more certain; and we invite you to follow us as we attempt to exhibit the coming struggles and future triumphs of the student of the skies.

LECTURE III.

THEORIES FOR THE EXPLANATION OF THE MOTIONS OF THE HEAVENLY BODIES.

IF in tracing the career of astronomy in the primitive ages of the world, we have been left to pursue our way dimly, through cloud and darkness,—if regrets rise up, that time has swept into oblivion the names and country of the early discoverers, in one reflection there is some compensation—while the bright and enduring truths which they wrested from nature have descended to us, their errors, whatever they may have been, are forever buried with their names and their persons. We are almost led to believe that those errors were few and transient, and that the mind, as yet undazzled by its triumphs, questioned nature, with that humility and quiet perseverance, which could bring no response but truth.

In pursuing the consequences flowing from the prediction of an eclipse, several remarkable results were reached, which we proceed to unfold. It will be recollected, that to produce either solar or lunar eclipses, the new or full moon must be in the act of crossing the sun's annual track. This point of crossing, called the *moon's node*, became therefore an object of the deepest interest. Long and careful scrutiny revealed the fact of its movement around the ecliptic, in a period of eighteen years and eleven days, during which time there occur 223 new moons, or 223 full moons. If then, a new moon falls on the sun's track to produce a solar eclipse to-day, at the expiration of 223 lunations, again will the new moon fall on the ecliptic, and an eclipse will surely take place. Suppose then that all the eclipses, which occur within this remarkable period of 223 lunations, are carefully observed, and the days on which they fall recorded, on each and every one of these days, during the next period of 223 lunations, eclipses may be expected, and their coming foretold.

This wonderful period of eighteen years and eleven days, or 223 lunations, was known to the Chaldeans, and by its use eclipses were predicted, more than 3000 years ago. It is likewise found among the Hindus, the Chinese, and the Egyptians, nations widely separated on the earth's surface, and suggesting the idea that it had its origin among a people even anterior to the Chaldeans. It is now known by the name of the Zaros, or Chaldean period.

Let it not be supposed that the application of the Zaros to the prediction of eclipses, can in any way supersede modern methods.—While antiquity contented itself with announcing the *day* on which the dark body of the moon should hide the sun, modern science points to the exact *second* on the dial, which shall mark the first delicate contact of the moon's dark edge with the brilliant disk of the sun.

It would be a matter of great interest to fix the epoch of primitive discovery. Though this is impossible, its high antiquity is attested by a few facts, to which we will briefly advert. We find among all the ancient nations, Chaldeans, Persians, Hindus, Chinese, and Egyptians, that the seven days of the week were in universal use, and what was far more remarkable, each of these nations named the days of the week after the seven planets, numbering the sun and moon among the planets. It is moreover found, that the order of naming is not that of the distance, velocity or brilliancy of the planets, and neither does the first day of the week coincide among the different nations; but the order once commenced is invariably preserved by all. If we compute the probability of such a coincidence resulting by accident, we shall find the chances millions to one against it. We are therefore forced to the conclusion, that the planets were discovered, and the seven days of the week devised and named, by some primitive nation, from whom the tradition descended imperfectly, to succeeding generations.

A remarkable discovery, made in the remote ages of the world, throws some farther light on the era of the primitive astronomical researches.— The release of the earth from the icy fetters of winter, the return of spring, and the revivification of nature, is a period hailed with uncommon delight, in all ages of the world. To be able to anticipate its coming, from some astronomical phenomenon, was an object of earnest investigation by the ancients.

It was found that the sun's entrance into the equinox, reducing to equality the length of the day and night, always heralded the coming of the spring. Hence to mark the equinoctial point among the fixed stars, and to note the place of some brilliant star whose appearance in the early morning dawn, would announce the sun's approach to the equator, was early accomplished with all possible accuracy. This star once selected, it was believed that it would remain forever in its place.

The sun's path among the fixed stars had been watched with success, and it seemed to remain absolutely unchanged, and hence the points in which it crossed the equator, for a long while were looked upon as fixed and immovable. And indeed centuries must pass away before any change could become sensible to the naked eye and its rude instrumental auxiliaries. But a time arrives at last when the bright star which for more than five hundred years had, with its morning ray announced the season of flowers, is lost. It has failed to give its warning—spring has come, the forests bud, the flowers bloom, but the star which once gave promise, and whose ray had been hailed with so much delight by many generations, is no longer found. The hoary patriarch recalls the long experience of a hundred years, and now perceives, that each succeeding spring had followed more and more rapidly after the appearance of the sentinel star. Each year the interval from the first appearance of the star in the early dawn, up to the equality of day and night, had grown less and less, and now the equinox came, but the star remained invisible, and did not emerge from the sun's beams until the equinox had passed.

Long and deeply were these facts pondered and weighed.—At length truth dawned, and the discovery broke upon the unwilling mind, that the sun's *path among the fixed stars was actually changing*, and that his point of crossing the equator was slowly moving backwards towards the west and leaving the stars behind. The same motion, only greatly more rapid, had been recognized in the shifting of the moon's node and in the rapid motion of the points at which her track crossed the equator. The retrograde motion of the equinoctial points, caused the sun to reach these points earlier than it would have done had they remained fixed, and hence arose the *precession* of the *equinoxes*.

This discovery justly ranks among the most important achieved by antiquity. Its explanation was infinitely above the reach of human effort at that early day: but to have detected the fact, and to have marked a motion so slow and shrouded, gives evidence of a closeness of observation worthy of the highest admiration. It will be seen hereafter, that the human mind has reached to a full knowledge of the causes producing the retrograde movement of the equinoxes among the stars. Its rate of motion has been determined, and its vast period of nearly twenty-six thousand years has been fixed. Once revealed, the slow movement of the equinox, makes it a fitting hour hand, on the dial of the heavens, with which to measure the revolutions of ages. As the sun's path has been divided into twelve constellations, each filling the twelfth part of the entire circuit of the heavens, for the equinox to pass the twelfth part of the dial, or from one constellation to the next, will require a period of more than two thousand years. Since the astronomer first noted the position of this hour hand on the dial of the stars, but one of its mighty hours of two thousand years, has rolled away. In case any record could be found, any chiselled block of granite, exhibiting the place of the equinox among the stars, at its date, no matter if ten thousand years had elapsed, we can reach back with certainty and fix the epoch of the record.

No such monument has ever been found; but there are occasional notices of astronomical phenomena, found among the Greek and Roman poets, which at least give color to conjecture. Virgil informs us that " the White Bull opens with his golden horns the year."

" Candidus auratis aperit cum cornibus annum,'
" Taurus."

This statement we know is not true, if applied to the age in which the poet wrote, and seems to be the quotation of an ancient tradition. If this conjecture be true, this tradition must have been carried down the stream of time for more than two thousand years, to reach the age in which the poet wrote. Although these conjectures are vague and uncertain, the frequent allusions to the constellations of the zodiac, in the old Hebrew Scriptures, and in the works of all ancient writers, sufficiently attest the extreme antiquity of these arbitrary groupings of the stars.

In taking leave of the primitive ages of astronomy and in entering on that portion of the career of research and discovery whose history has been preserved, let us pause for a moment and consider the position occupied by the human mind at this remarkable epoch.

Thus far the eye had done its work faithfully. Through long and patient watching, it had revealed the facts, from which reason had wrought out her great results. The stars grouped into constellations glittered in the blue concave of a mighty sphere, whose centre was occupied by the earth. Within this hollow sphere, sun, moon and planets, kept their appointed courses, and performed their ceaseless journeys. Their wanderings had been traced,—their pathway in the heavens was known,—their periods determined,—the inclinations of their orbits fixed. So accurately had the eye followed the sun and moon, that it had learned to anticipate their relative positions, their oppositions and conjunctions, till reaching forward, it had robbed the dread eclipse of its terrors, and had learned to hail its coming with delight. The pathway of the sun and moon among the stars had been scanned and studied, until their slowest changes had been marked and measured.

Such were the rich fruits of diligence and perseverance which descended from the remote nations of antiquity. With the advantage of these great discoveries, and the experience of preceding ages, it is natural to expect rapid progress, when science found its home among the bold, subtle, and inquisitive Greeks. He who entertains this expectation will meet with disappointment. Not that investigations were less constantly or perseveringly conducted,—not that less perfect means were employed, or less powerful talent consecrated to the work; but because a point had been reached of exceeding difficulty. The era of discovery from mere inspection was rapidly drawing to a close. It was an easy matter to count the days from full moon to full moon, to watch a planet as it circled the heavens, from a fixed star until it returned to the same star again, to mark its stopping, its reversed motion, and its onward goings; but it was a far different matter to rise to a knowledge of the causes of these stations and retrogradations, and to render a clear and satisfactory account of them. The problem now presented, was to combine all the facts treasured by antiquity, all the movements exhibited in the heavens, and reduce them to simplicity and harmony. The Greek philosophers, from Plato down to the extinction of the last school of philosophy, recognized this to be the true problem, and essayed its solution, with an energy and pertinacity worthy of the highest admiration.

Let us now examine the causes which arrested the progress of astronomical discovery and held back the mind for a period of more than two thousand years. Surrounded as we are by the full blaze of truth, accustomed to the simplicity and beauty which now reign everywhere in the heavens, we find it next to impossible to realize the true position of those brave minds, which, enveloped in darkness, deceived by the senses, fettered by prejudice, struggled on and finally won the victory, whose fruits we enjoy.

The most careful and philosophical examination of the heavens seemed to lead to the admitted truth, that the earth was the centre of all celestial motion. In the configuration of the bright stars there was no change. From age to age, from century to century, immovably fixed in their relative positions, they had performed their diurnal revolutions around the earth. They were ever of the same magnitude, of the same brilliancy. How impossible was this, on any hypothesis, except that of fixed central position of the earth. Leaving the fixed stars, an examination of the motions of the sun and moon—their nearly uniform velocity—their invariable diameters in all portions of their orbits, demonstrated the central position of the earth with reference to them. To shake a faith thus firmly fixed, sustained by the evidence of the senses, consonant with every feeling of the mind, accordant with fact and reason, required a depth of research, and the development of new truths, only to be revealed after centuries of observation.

Every effort, then, to explain the celestial phenomena, started with the undoubted fact, that the earth was the centre of all motion. Thus far, the mind had not reached the idea of apparent motion. If the moon moved, so equally did the sun. There was exactly the same amount of evidence to demonstrate the reality of the one motion, as the other; neither were doubted. It would have been unphilosophical to reject the one, without rejecting the other.

The centre of motion once determined, the nature of the curve described was so obviously presented to the eye, that it seemed impossible to hesitate for one moment. The circle was the only regular curve known to the ancients. Its simplicity, its beauty and perfection, would have induced its selection, even had there been a multitude of curves from which to choose. Its curvature was ever the same. It had neither beginning nor end. It was the symbol of eternity, and admirably shadowed forth the eternity of the motions to which it gave form. As if these considerations had required confirmation, every star and planet, the sun and moon, all described circles, in their diurnal revolution, and it seemed impossible to doubt that their orbitual motions were performed in the same beautiful curve. In truth, observation confirmed this conjecture; and the orbits of all the moving bodies, when projected on the concave heavens, were circles. That this curve, then, should have been adopted without doubt or hesitation, is not to be wondered at. It came therefore, to be a fixed principle, that in all hypotheses devised to explain the phenomena of the heavens, circular motion and circular orbits, alone could be employed.

To these great principles, of the central position of the earth, and the circular orbits, we must add that of the earth's immobility. This doctrine was undoubtedly sustained by the evidence of all the senses which could give testimony. No one had seen it move,—had heard it move,—had felt it move. How was it possible to doubt the evidence of the eye, the touch, the ear? Here, then, was another incontrovertible fact, which even the most skeptical could not doubt, and which laid at the foundation of all effort to resolve the problem under examination.

With a full knowledge and appreciation of these facts, we are prepared to enter upon an examination of the career of astronomy, up to the time when all darkness disappeared before the dawning of a day which should never end. The early Greek philosophers, little fitted by nature for close and laborious observation, rather chose to gather in travel the wisdom which was garnered up in the temples, and among the priests of Egypt, and India. Returning to their native country, they theorized on the facts they had learned, and taught doctrines, which found their only support in trains of fanciful or specious reasoning. Thus we find Pythagoras mingling the great discoveries of antiquity with theories the most vague and visionary. While gleams of truth flash occasionally through the darkness of his doctrines, they seem but fortunate guesses. His views were sustained by no solid argument, and rapidly sunk into forgetfulness. This philosopher is said to have fixed the sun in the centre of his planetary system, and to have taught the revolution of the earth in an orbit; but to sustain this bold conjecture, the only reason assigned, was, that fire which composes the sun, was more dignified than earth, and hence should hold the more dignified position in the centre. We are not surprised that Hipparchus and Ptolemy, the true astronomers among the Greeks, should have rejected a doctrine sustained by so futile and absurd a reason. Nicetas, a follower of Pythagoras, is said to have gone farther than his master, and to have adopted the idea that the revolution of the heavens, was an appearance produced by an actual rotation of the earth on an axis, once in twenty-four hours. This extraordinary and almost prophetic announcement, unfortunately was not sustained by any solid argument. It was regarded as a vain dream, and soon was lost in oblivion.

A crowd of theoretic philosophers filled for a long time the school of Greece, contributing little to science, and diverting the mind from the only train of research which could lead to any true results. At length a philosopher arose who restored investigation to its legitimate channel. Hipparchus, abandoning, for the present, all vain effort to explain the phenomena of the heavens, gave himself up to close, continuous and accurate observation. He began with the movements of the sun in his annual orbit. By the construction of superior brazen circles, he measured the daily motion of the sun during the entire year. He confirmed the discovery of the ancients, of the irregular or unequal progress of this luminary, and fixed that point in the sun's orbit where it moved with greatest velocity. Year after year, did this devoted astronomer follow the sun, until finally he discovered that the point on the orbit, where its motion was swiftest, did not remain fixed, but was advancing in each revolution, at a very slow rate along the orbit. Having thus demonstrated and characterized the irregularity of the sun's motion, he directed his attention to minute examinations of the moon, and reached results precisely similar. From these discoveries, it became manifest, that in case the motions of the sun and moon were circular and uniform, the earth did not occupy the exact centres of their orbits; for on this hypothesis any irregularity of motion would have been impossible. Here was a point gained. The exact central position of the earth was disproved in two instances, and even the amount of its eccentricity, or distance from the true centre, determined. Retaining the circular and uniform motion of the sun and moon, the discovered irregularities were tolerably well represented by the eccentric position of the earth, from whose surface these motions were measured.

While pursuing these important researches, Hipparchus resolved upon a work of extraordinary difficulty, which had never before been attempted and which fully attests the grandeur and sagacity of his views. This enterprise was nothing less than numbering the stars and fixing their positions in the heavens. This he actually accomplished, and his catalogue of 1081 of the principal stars, is perhaps the richest treasure which the Greek school has transmitted to posterity. We cannot too much admire the disinterested devotion to science, which prompted this great undertaking, and the firmness of purpose which sustained the solitary observer, through long years of toil. It was a work for posterity, and could yield to its author no reward during his life. Conscious of this, his resolution never faltered, and grateful posterity crowns his memory with the well-earned title of Father of Astronomy. The noble example thus set by Hipparchus, was not lost on Ptolemy, justly the most distinguished among his immediate successors. An ardent student, a close observer, a patient and candid reasoner, Ptolemy collected and digested the discoveries and theories of his predecessors, and transmitted them, in connection with his own, successfully to posterity. Rejecting the absurd doctrine of the solid crystal spheres of Eudoxus, and the unsustained notions of Pythagoras, this bold Greek undertook the resolution of the great problem which Plato had long before presented, and to accomplish which, so many unsuccessful efforts had been made.

After a careful examination of all the facts and discoveries, which the world then possessed, adding his own extensive observations, Ptolemy promulged a system which bears his name, and which endured for more than fourteen hundred years. He fixed the earth as the great centre, about which the sun, the moon, the planets, and the starry heavens revolved. Retaining the doctrine of uniform circular motion, he accounted for the irregularity in the movements of the sun and moon by the eccentric position of the earth in their orbits.—To explain the anomalous movement of the planets, he devised the system of cycles and epicycles. Every planet moved uniformly in the circumference of a small circle, whose centre moved uniformly in the circumference of a large circle, near whose centre the earth was located. By this ingenious theory, it was shown that a planet moving in the circumference of its small circle might appear to retrograde, to become stationary, and finally to advance among the fixed stars. Thus were all the phenomena known to the Greek astronomer, so satisfactorily accounted for, that it even became possible from this singular theory, to compute tables of the planetary motions, from which their places could be predicted with such precision, that the error if any existed, escaped detection by the rude instruments then in use.

While the explanation of the celestial phenomena had constituted the principal object of the Greek astronomers, some rude efforts were commenced to determined the magnitude of the earth, and the relative distances of the sun and moon. The process adopted by Eratosthenes, two thousand years ago, to determine the circumference of the earth, and its diameter, is essentially the same now employed by modern science. The results reached by the Greek astronomer, owing to an ignorance of the exact value of his unit, are lost to the world.

When astronomy was banished from Greece, it found a home among the Arabs. When darkness and gloom wrapped the earth through ten long centuries, and human knowledge languished, and art died, and genius slumbered, it is a remarkable fact, that astronomy during that long period of ignorance, instead of being lost, was actually slowly advancing, and when the dawn of learning once more broke on Europe, the astronomy of the Greeks, improved by the Arabs and the Persians, was preserved in the great work of Ptolemy, and transmitted to posterity.

It is true that no change had been wrought in the Greek theory, but observations had been multiplied and slow changes measured, which prepared the way for the discoveries which were soon to succeed. On the revival of learning in Europe, the literature and science of the Greeks and Romans rapidly spread, and gained an astonishing ascendancy over the human mind. Indeed, theirs was the only science, the only wisdom. Time honored, and venerable with age, the philosophy of Aristotle, the geometry of Euclid, and the astronomy of Ptolemy, filled the colleges and universities, and fastened itself upon the age, with a tenacity, which permitted no one to question or doubt, and which seemed to defy all further progress.—Such was the state of science and the world, when Copernicus consecrated his genius to the examination of the heavens.

To a mind singularly bold and penetrating, Copernicus united habits of profound study and severe observation. Deeply read in the received doctrines of science, he examined with the keenest interest, every hint which the philosophers of antiquity had left on record concerning the system of nature. For more than thirty years he watched, with unceasing perseverance, the movements of the heavenly bodies. By the construction of superior instruments, he compared the observed places of the sun, moon and planets, with their positions computed from the best tables founded on the theory of Ptolemy. The hypothesis of uniform circular motion, had originally been adopted, to preserve the simplicity of nature and with true philosophy. But as one irregularity after another had been discovered in the movements of the heavenly bodies, each of which must be explained on the circular hypothesis, one circle had been successively added to another, eccentrics and epicycles, equants and differents, until to preserve simplicity, the system had grown to the most extravagant complexity. The primitive idea of simplicity was a just one, founded in nature, and adopted in reason. But after thirty years of vain effort to harmonize the phenomena of the heavens with the theory of Ptolemy, after entangling himself in a maze of complexity in his effort to preserve simplicity, Copernicus was at last driven to doubt, and doubt soon grew into disbelief. By a close examination of the motions of Mercury and Venus, he found that these planets always accompanied the sun, participated in its movements, and never receded from it except to limited distances. The uniformity of their oscillations, from the one side to the other of the sun, suggested their revolution about that luminary, in orbits, whose planes passed nearly through the eye of the observer. The Egyptians had reached to this doctrine, had communicated it to Pythagoras, who taught it to his countrymen, nearly two thousand years before the time of Copernicus.

If then simplicity imperiously demanded the abandonment of the earth as the great centre of motion, in the search for a new centre, a multitude of circumstances pointed to the sun. It was the largest and most brilliant of all the heavenly bodies. It gave light to the moon and planets. It gave life to the earth and its inhabitants. It was certainly accompanied by two satellites, and above all, it was so related to the earth, that if motion in the one was abandoned, it must instantly and without a moment's hesitation, be transferred to the other. Long did the philosopher hesitate, perplexed with doubts, surrounded by prejudice, embarrassed with difficulties, but finally rising superior to every consideration save truth, he quitted the earth, swept boldly through space, and planted himself upon the sun. With an imagination endowed with the most extraordinary tenacity, he carried with him all the phenomena of the heavens, which were so familiar to his eye, while viewed from the earth. A long train of investigation was now before him. He commences with his now distant earth. Its immobility is gone—he beholds it sweeping round the heavens in the precise track once followed by the sun. The same constellations mark its career, the same periodic time, the same inequalities of motion: all that the sun has lost the earth has gained.

Thus far the change had been without results. He now gives his attention to the planets. Here a most beautiful scene broke upon his senses. The complex wanderings of the planets, their stations, their retrograde motions, all disappeared, and he beheld them sweeping harmoniously around him. The earth, deprived of her immobility, started in her orbit, joined her sister planets, and gave perfection to the system. The oscillations of Mercury and Venus were converted into regular revolutions, still holding their places nearest to the sun; then came the earth, next Mars, and Jupiter, and last of all Saturn away in the distance, slowly pursuing his mighty orbit. All were moving in the same direction, their paths filling the same belt of the heavens.

Charmed with this beautiful scene, the philosopher turns to an examination of the moon. Was she, too, destined to take her place among the planets. A short investigation revealed her true character. She could not be a planet revolving about the sun *interior* to the earth's orbit, for if so she would have imitated the oscillations of Mercury and Venus. She was not a planet revolving around the sun, *exterior* to the orbit of the earth, for in that case she must have imitated the stations and retrogradations of Mars, Jupiter, and Saturn. The invariability of her diameter as seen from the earth, joined to these considerations, established the fact of her secondary character, and like a favorite minister who accompanies his dethroned monarch in his exile, so did the faithful moon cling to the earth and follow it in its wanderings through space.

Such is the beautiful system wrought out by the great Polish philosopher. Far from perfect, it was founded in truth, and although improvement might and must come, revolution could never shake its firm foundation.— While the more prominent irregularities in the planetary motion, were removed by constituting the sun the centre of motion, there yet remained an increase and decrease in the orbital velocities of all the planets, now including the earth among the number, which were inexplicable. The planets did not revolve, then, in circles whose exact centre was occupied by the sun. The moon's orbit was not a circle, whose exact centre was the earth; and to explain these unfortunate irregularities, Copernicus clinging to circular motion, as the world had done for 2000 years, was driven to adopt the same expedients which characterized the theories of

Ptolemy: the eccentric and epicycle were fastened upon the new system of astronomy. Yet another difficulty embarrassed the mind of Copernicus. In giving to the earth a rotation on its axis once in twenty-four hours, he explained the apparent revolution of the starry heavens. This axis of rotation, it was readily seen, must ever remain parallel to itself in the annual revolution of the earth in its orbit. Being in this way carried round such a vast circumference, the prolongation of the axis ought to pierce the northern heavens in a series of points which would form a curve so large as not to escape detection. But no such curve appeared, the north pole of the heavens, scrutinized with the most delicate instruments, preserved its position, immovably throughout the entire revolution of the earth in its orbit, and to escape from this difficulty there was no alternative but to admit that the distance of the sphere of the fixed stars was so great that the diameter of the earth's orbit, equal to 200,000,000 of miles was absolutely nothing, when compared with that mighty distance.

Under these circumstances, it is not wonderful that Copernicus should have promulgated his system with extreme diffidence and only after long delay; indeed his great work, setting forth his doctrines, was never read by its author in print, and only reached him in time to cheer his dying moments.

We cannot then be surprised, that the new system was received with doubt and distrust, or rather that it was for a long while absolutely rejected.—The progress of truth is ever slow, while error moves with rapid pace. The reason is obvious.—Error is seized by a class of minds, which asks no evidence; while the searchers for truth, adopt it only after the most deliberate examination.

But the revolution had been commenced. A few bold minds were struck with the simplicity and beauty of the conjectures of Copernicus; and when the exigencies of the age demand genius, it seems to rise spontaneously. The mind had persevered in a system founded in reason, and which nothing short of this very perseverance could have demonstrated to be erroneous. Like the traveler, who is uncertain which of two roads to take, he reflects, reasons, and decides, and even if his choice be a wrong one it would be folly to stop before fully convinced that he had chosen erroneously.

But the mind is once again in the path of truth; and after wandering twenty long centuries in darkness, which grew deeper and deeper, the change from darkness to light gives vigor to its movements, and its future achievements are destined to be rapid and glorious.

Here let us pause for a moment, on the boundary which divides ancient from modern science, and glance at the collateral circumstances which were found to modify and retard the investigations which had commenced. The old doctrines of philosophy and astronomy, had become intimately interwoven with human society. Ptolemy, and Plato, and Aristotle, were regarded with a sort of reverential awe. Even the church, not following, but leading the world in this profound respect for ancient philosophy, pronounced the doctrines of Ptolemy in accordance with the revelations of scripture, and girdled them with the fires of persecution, through which alone their sacredness could be attacked. Thus entrenched, and defended by prejudice, by society and by religion, none but the most daring spirit would enter the conflict against such unequal odds. Conscious of these difficulties, Copernicus had wisely avoided collision, and gave his doctrines to the world with such caution as not to provoke attack. But this armed neutrality could not long endure. If the new doctrine were founded in error, left to itself it would never advance, and would soon quietly sink into oblivion. On the contrary, should it prove to be based upon truth, no power could arrest its progress, or stay its development. The contest must come sooner or later, and demanded in those who should battle for the truth the rarest qualities.

Copernicus had merely commenced the examination of his bold conjecture. A life-time was too short to accomplish more. He had transferred the center of motion from the earth to the sun, and rested the truth of his hypothesis on a *diminished* complexity in the celestial phenomena. In case the true centre had been found, it now remained to determine the exact curves in which the planets revolved, the laws regulating their motion, and the nature of the bond which it was now suspected, united the planetary worlds into one great a system. The resolution of these profound questions was reserved for *Kepler*, who has without flattery been termed the legislator of the heavens, and who has earned the reputation of being *first* in fact and first in genius among modern astronomers. He united in the most perfect manner, all the qualifications of a great discoverer. Ardent, enthusiastic, and subtle, he pursued his investigations with a keen and restless activity. Patient, laborious, and determined, difficulties shrunk at his approach, and obstacles melted before him. Unprejudiced and pious, he sought for truth in the name and invoking ever the guidance of the great Author of truth. If his theories were not actually deduced from facts, when formed, no test was too severe, and nothing short of a rigid coincidence with fact could satisfy the exacting mind of this wonderful genius. Realizing fully the difficulty and importance of the researches before him, once commenced, his perseverance knew no limit, and the fertility of his imagination was utterly inexhaustible.

Such was the man to whom the interests of science at this critical juncture were committed. Having adopted as an hypothesis, the central position of the sun, and the revolution of the earth and planets around this centre, he determined to discover the true nature of the planetary orbits, and find if possible, some single curve which would explain the orbital motions of the celestial bodies. To accomplish this difficult enterprise,

Kepler wisely determined to confine his efforts and investigations to one single planet, and Mars was selected as the subject for experiment. He commenced by a rigorous comparison between the observed places of the planet and those given by the best tables which could be computed by the circular theory. Sometimes the predicated and observed places agreed well with each other, and hope whispered that the true theory had been found; but pursuing the planet onward in its sweep around the sun, it would begin to diverge from its theoretic track, its distance would increase until it became evident that the theory was false, and must be abandoned.

Nothing daunted, the ardent philosopher consoled himself with the thought, that among all possible theories which the mind could frame, one had been stricken from the list, and a diminished number remained for examination. This was a new mode of research, and in case the number of theories was not too great, and the patience of the philosopher sufficiently enduring, a time would come, sooner or later, when success must reward his labors. Thus did Kepler toil on subjecting one hypothesis after another to the ordeal of rigid experiment, until no less than nineteen had been tested with the utmost severity and all were rejected. Eight years of incessant labor had been devoted to this examination. He had exhausted every combination of circular motion which the fertility of his imagination could suggest. They had all utterly failed.—The charm was ended, and he finally broke away from the fascination of this beautiful curve, which for five thousand years had so bewildered the human mind, and boldly pronounced it impossible to explain the planetary motions with *any* circular hypothesis. This at least was a great negative triumph. If he had not found the curve in which the planets revolved, he had found what it could *not* be, and released from all future embarrassment from eccentrics and epicycles, he now pursued a lofty and independent train of investigation.

Leaving forever the circle, the next simplest curve is called the ellipse, an oval figure, which when but little flattened very nearly resembles the circle in form but enjoys very different properties. All diameters of a circle are equal. The diameters of an ellipse are unequal. The centre of the circle is equally distant from all points on the circumference. No such point exists in the ellipse; but two curious points are found on its longest diameter, possessing the remarkable property of having the sum of the lines joining them with any point of the curve constantly equal to the longest diameter. Each of these points is called a *focus*. This beautiful curve, with its singular properties, had been discovered by the Greek mathematicians; but not remarking its use in nature, it had hitherto been regarded only as an object of amusing speculation. To this curve did Kepler apply, when driven from the circular hypothesis, and again commenced his system of forming hypotheses, and hunting them down, as he termed his scrutinizing process. As in the circular hypothesis the sun had at first been located in the centre, so in commencing the elliptic theory the centre of the longest diameter was made the centre of motion. Buoyant with hope, the astronomer sets out to follow the planet around its elliptic orbit; but although for a short distance its movements were well represented, it finally broke away from the elliptic track, and bid defiance to the central hypothesis. But Kepler was not in the least disheartened with this first effort.—He now shifts the sun to the focus of the ellipse, constructs his orbit, starts once more on the track of the planet, watches it as it sweeps onward around the sun, the elliptic orbit holds it as it moves farther and still farther.—Half its revolution is performed and there is no diverging, onward it flies,—the goal is won.— Triumph crowns the philosopher, *the orbit is found!*

Thus was accomplished one of the most important discoveries which the mind had ever reached. The elliptic orbit of Mars rapidly led to those of the other planets, and to that of the moon, and Kepler proclaimed to the world his first great law, in the following language: *Planets revolve in elliptic orbits about the sun, which occupies the common focus of all these orbits.*

This law swept forever from the heavens and from astronomy those complications which had stood the test of centuries, nay of thousands of years. Their mysterious power was paralyzed by this single touch of the enchanter's wand, and they fled from the skies. The circle was as simple and beautiful as ever, but its divine character was gone, and the gods or angels who had so long held their abodes in the planets were exiled from their homes. The dawn of *modern* science broke in beauty on the world.

Kepler having been so signally rewarded by this great discovery, now turned his attention to an investigation of the first importance, one indeed which was indispensably necessary to render his first discovery available. As the planets were now known to revolve in ellipses, and as their motion was found by observation to be unequal in different parts of their orbits, it became a matter of the first consequence to ascertain some simple law regulating the orbitual motion, and by means of which a planet might be readily followed, and its places computed. To detect this law, in whose existence Kepler seems to have entertained the most unwavering faith, a figure was drawn representing the orbit of Mars, the sun occupying one of the foci of the curve. On the circumference of this curve the places of the planets were marked down as observation had determined them; and here commenced a series of examinations which finally led to the knowledge of the second great law of the planetary motions, which may be thus announced. *If a line be drawn from the centre of the sun, to any planet, this line as it is carried forward by the planet will sweep over equal areas in equal portions of time.*—This law accorded in the most perfect manner with fact, and gave at once the power of following, and from the mean motion, computing the place of any planet; a triumph which all the complexity of older systems had failed ever to accomplish.

Any other mind less adventurous than that of Kepler, might have been satisfied with these two great discoveries. The precise curves described by the planets and a law regulating their motions in their orbits, sufficed to render all the phenomena of the heavenly bodies not only explicable, but susceptible of accurate prediction.—There seemed nothing more to be added.—Kepler did not think so. He conceived the idea that the solar system was not a mere assemblage of isolated planets revolving about a common centre, but a great associated system, in which some common bond of union existed, which once found, would present the solar system in a new and true light.

This bond he believed existed in some hidden relation between the times occupied by the planets in describing their orbits, and their distances from the sun. In the history of this remarkable research, we are presented with one of the brightest examples of the fruits of perseverance. If some superior power, some spirit from a brighter world, had revealed to the mind of Kepler the actual existence of some relation between the planet's periods and distances, and had proposed to him to discover this hidden law, there would have been a definite object before the astronomer, and to have persevered in the pursuit of this object, would have been within the limits of probability, even if a lifetime were exhausted in fruitless efforts. But to excite in his own mind a faith sufficiently strong in the existence of a law of which there existed not the slightest evidence, and to have persevered in its research for seventeen long years of laborious effort, seems almost incredible.

There is an immense difference between the pursuit which resulted in the discovery of the first two laws of Kepler, and the third one. In seeking for the curve described by the planets, it was looking for that, which must have an existence; and in tracing the law of a planet's motion, it was absolutely impossible to follow a planet, or predict its positions, without such a law. But in seeking for a bond of union among the planetary periods and distances, it was a search for that, which it was believed had no existence, except in the wild imagination of this extraordinary philosopher. The history of mind scarcely furnishes an example in any degree paralleled, if we except perhaps the heroic fortitude which marked the career of Columbus.—Yet even the great Genoese was in possession of solid facts on which to base his reasoning. He saw evidences of the existence of another hemisphere, which the superficial could never realize. Kepler, more bold, more grand, more sublime, dreamed of nothing less than a brotherhood of worlds, a mighty and magnificent scheme of vast revolving orbs. Should success crown his efforts, the most brilliant results would follow. The distance of a single planet from the sun once obtained, and the periodic time of all being known, the distances might then be found for each individual in the entire system, without even directing an instrument to the heavens. Here then was a prize to reach which no time, or pains, or labor could be misapplied. Its return would be a hundred-fold.

But where was the prize to be sought? Even admitting that some common bond did bind the circling worlds into one harmonious system, did it exist in some hidden relation between their periods of revolutions, their distances, their magnitudes, their densities? or was it to be sought in some analogy between the distances and periodic times? After long and deliberately pondering this great problem, Kepler decided that the strongest probability suggested that the distances of the planets and their periods of revolution, would in some way contain the mysterious bond of union. Here then did this daring mind concentrate its energies; and his purpose once fixed, he marches steadily forward in his research with a courage which no defeat could daunt, and a perseverance which knew no limit but success.

Before announcing the final result, let me explain two terms employed in its statement. The *square* of any quantity results by multiplying it by itself. The *cube* comes from multiplying the *square* by the number. The square of a planet's period, or the cube of its distance, are known the moment we know the period, and distance, by applying the simple rules of arithmetic. After Kepler had exhausted all simple relations between the periods and distances of the planets, in no degree shaken in his lofty faith, he proceeded to try all possible relations between the squares of the periods and distances; but with as little success. Nothing daunted, he proceeded to investigate the possible relations between the cubes of the periods and distances. Here again he was foiled; no law exhibited itself.—He returned ever fresh to the attack, and now commenced a series of trials involving the relations between the simple periods and the squares of the distances. Here a ray of hope broke in upon his dim and darkened path.

No actual relation existed, yet there was a very distant approixmation, enough to excite hope.—He then tried simple multiples of the periods and the squares of the distances—all in vain. He finally abandoned the simple periods and distances, and rose to an examination of the relations between the squares of these same quantities.—Gaining nothing here, he rose still higher, to the cubes of the periods and distances;—no success, until finally he tried the proportion existing between the squares of the periods in which the planets perform their revolutions and the cubes of their distances from the sun.—Here was the grand secret, but, alas! in making his numerical computations, an error in the work vitiated the results and with the greatest discovery which the mind ever achieved in his very grasp, the heart-sick and toil-worn philosopher turned away almost in despair from his endless research.

Months rolled round, and yet his mind with a sort of keen instinct, would recur again and again to this last hypothesis. Guided by some kind angel or spirit whose sympathy had been touched by the unwearied zeal of the mortal, he returned to his former computations, and with a heaving breast, and throbbing heart, he detects the numerical error in his work, and commences anew. The square of Jupiter's period is to the square of Saturn's period as the cube of Jupiter's distance is to some fourth term, which Kepler hoped and prayed might prove to be the cube of Saturn's distance. With trembling hand, he sweeps through the maze of figures; the fourth term is obtained; he compares it with the cube of Saturn's distance.—They are the same!—He could scarcely believe his own senses. He feared some demon mocked him.—He ran over the work again and again.—He tried the proportion, the square of Jupiter's period to the square of Mars' period as the cube of Jupiter's distance to a fourth term, which he found to be the cube of the distance of Mars.—Till finally full conviction burst upon his mind: he had won the goal, the struggle of seventeen long years was ended, God was vindicated, and the philosopher in the wild excitement of his glorious triumph, exclaims: "Nothing holds me. I will indulge my sacred fury! If you forgive me I rejoice; if you are angry I can bear it. The die is cast. The book is written, to be read either now, or by posterity, I care not which. —It may well wait a century for a reader, since God has waited six thousand years for an observer!"

More than two hundred years have rolled away since Kepler announced his great discoveries. Science has marched forward with swift and resistless energy. The secrets of the universe have been yielded up under the inquisitorial investigations of god-like intellect. The domain of the mind has been extended wider and wider. One planet after another has been added to our system; even the profound abyss which separates us ⁻om the fixed stars has been passed, and thousands of rolling suns have een described, swiftly flying or majestically sweeping through the hronged regions of space. But the laws of Kepler bind them all,— satellite and primary—planet and ʀun—sun and system—all with one accord, proclaim in silent majesty, the triumph of the hero philosopher.

LECTURE IV.

DISCOVERY OF THE GREAT LAWS OF MOTION AND GRAVITATION.

THE remarkable discoveries which had rewarded the researches of Kepler, confirmed in the most perfect manner the doctrines of Copernicus, flowing as they did from his prominent hypothesis, the central position of the sun. Having reached to the true laws of the planetary motions, the whole current of astronomical research was changed. New methods were demanded, and more delicate means of observation must be brought into use before the data could be furnished for new discoveries. Henceforward astronomy could only advance by the aid of kindred sciences. Mathematics, optics, and above all, mechanical philosophy, were to become the instruments of future conquests.

The philosophy of Aristotle, though very far from deserving it, wielded quite as extensive an influence over the age, as did the astronomy of Ptolemy. It appears, indeed, that the followers of Aristotle regarded their master as absolutely infallible, and gave to his doctrines a credence so firm, that even the clearest experiments, the most undeniable evidence of the senses were sooner to be doubted than the doctrines of the divine Greek. To attack and destroy a system so deeply rooted in the prejudices of the age, required a mind of extraordinary courage and power, a mind deeply imbued with the love of truth, quick in its perceptions, logical in argument, and firm in the hour of trial.

Such a mind was that of the great Florentine philosopher, Galileo Galilei, the senior, friend, and contemporary of Kepler. Indeed the exigencies of the age seemed to have given birth to three men, whose peculiar constitutions fitted them for separate spheres, each of the highest order, each in some measure independent, and yet all combining in the accomplishment of the great scientific revolution. While Tycho, the noble Dane, immured within the narrow limits of his little island, watching from his sentinel towers the motions of the stars, noting with patient and laborious continuity, the revolutions of the sun, moon and planets, was accumulating the materials which were to furnish the keen and inquisitive mind of Kepler with the means of achieving his great triumphs—Galileo, with a giant hand, was shaking to their foundations the philosophical theories of Aristotle, and startling the world with his grand mechanical discoveries. But for the observations of Tycho, Kepler's laws could not have been revealed;—but for the magic tube of Galileo, these laws had been the *ns plus* of astronomical science. Thus do we witness the rare spectacle of three exalted intellects, contemporaneously putting forth their diverse talents in the accomplishment of one grand object. The Dane, the German, and the Italian, divided by language and by country, united in the pursuit of science and of truth.

Called to Pisa to discharge the duties of a philosophical teacher, Galileo was not long in detecting the extravagant philosophical errors of Aristotle, which had been implicitly received for more than twenty centuries. He continued to teach the text of his old master, but it was only to expose its unsound and false philosophy to his wondering and incredulous pupils. A desecration so monstrous, could not long escape exposure and punishment. Indeed the Florentine made no secret of his teachings. The Aristotelians made common cause against the young philosophical heretic, and he was warned to desist from his heresy. Galileo gave for answer to his opponents, that he was ready to relinquish his new views the moment they were shown by experiment to be false; on the other hand, he demanded of them equal candor, and proposed to refer the matter in controversy to the tribunal of experiment.

Aristotle, in discussing the laws of falling bodies, affirmed the principle, that the velocity acquired by any falling body, was in the direct proportion of its weight; and if two bodies of unequal weight commenced their descent from the same height, at the same moment, the heavier would move as many times swifter than the lighter, as its weight exceeded that of the smaller body. Galileo doubted the truth of this principle, and on subjecting it to the test of experiment, he saw instantly that its variation from fact was wide as it could be. The obvious character of this experiment, its freedom from all chances of deception, and the importance of the principle involved, induced the young philosopher to select it as the test, and to challenge his opponents to a public demonstration of the truth or falsehood of their old system of philosophy.—The challenge was accepted. The leaning tower of Pisa presented the most convenient position for the performance of these experiments, on which Galileo so confidently relied for triumphant demonstration of the error of Aristotle; and thither on the appointed day the disputants repaired, each party perhaps with equal confidence. It was a great crisis in the history of human knowledge. On the one side, stood the assembled wisdom of the universities, revered for age and for science, venerable, dignified, united and commanding. Around them thronged the multitude, and about them clustered the associations of centuries. On the other, there stood an obscure young man, with no retinue of followers, without reputation or influence, or station. But his courage was equal to the occasion; confident in the power of truth, his form is erect, and his eye sparkles with excitement.

LAWS OF MOTION AND GRAVITATION.

But the hour of trial arrives. The balls to be employed in the experiments are carefully weighed and scrutinized to detect deception. The parties are satisfied. The one ball is exactly twice the weight of the other. The followers of Aristotle maintain that when the balls are dropped from the top of the tower, the heavy one will reach the ground in exactly half the time employed by the lighter ball. Galileo asserts that the weights of the balls do not affect their velocities, and that the times of descent will be equal; and here the disputants join issue.—The balls are conveyed to the summit of the lofty tower. The crowd assembles round the base—the signal is given—the balls are dropped at the same instant, and swift descending, at the same moment they strike the earth. Again and again is the experiment repeated, with uniform results. Galileo's triumph was complete. Not a shadow of doubt remained; but far from receiving, as he had hoped, the warm congratulations of honest conviction—private interest, the loss of place, and the mortification of confessing false teaching, proved too strong for the candor of his adversaries.—They clung to their former opinions with the tenacity of despair, and assailed the now proud and haughty Galileo with the bitter feelings of disappointment and hate.

The war was now openly declared, and waged with a fierceness which seems to have excited the mind of the young philosopher to the most extraordinary efforts. Driven from Pisa, by the numbers and influence of his enemies, no suffering or danger could drive from his mind the great truths which his researches by experiment were constantly revealing. His spirit was unbroken, and in retiring from the unequal contest, he hurled back defiance into the face of his conquered, though triumphant persecutors.

The mechanical investigations of Galileo, conducted with clearness and precision, soon led to the most important discoveries. He detected the law of falling bodies, and showed that the spaces described were proportional to the squares of the times; that is, if a body fell ten feet in one second of time, it would fall four times as far in two seconds, nine times as far in three seconds, and so on for any number of seconds. He studied with success the subject of the composition of forces, and demonstrated this remarkable proposition, which lies at the very foundation of all modern mechanical philosophy. It may be thus stated. If a body receive an impulse, which singly would cause it to move thirty feet in a second, on the line of the direction of the impulse, and at the same instant another impulse be communicated in a different direction from the first, and which if acting alone would cause the body to move on the line of direction of the second impulse forty feet in one second, under the joint action of these two impulses the body will move in a direction easily determined from those of the impulsive forces, and will fly with a velocity of fifty feet in the first second of time.

STRUCTURE OF THE UNIVERSE.

Such is the universal prevalence of this beautiful proposition, that no falling, flying, or moving body, whether it be the rifle ball, the cannon shot, or the circling planet, is free from its imperious sway. Strike the knowledge of this great truth from existence, and the magnificent structure which modern science has reared, falls in ruins at a single blow. It is founded in the simple but invariable laws of motion, and while these endure, this elegant discovery of the Florentine philosopher will remain as a monument to his sagacity and penetration.

Possessed of such rare qualities for philosophic research, so free from prejudice, and withal, so candid, we cannot but inquire with interest, how the mind of Galileo stood affected towards the new astronomical doctrines of Copernicus. He had early adopted and taught the Ptolemaic system, and his conversion is so remarkable, and. is so characteristic of the man, that it cannot be omitted. A disciple of Copernicus visited the city of Galileo's residence, and delivered several public lectures to crowded audiences, on the new doctrines. Galileo, regarding the whole subject as a species of solemn folly, would not attend. Subsequently, however, in conversing with one who had adopted these new doctrines, the Copernican sustained his views with such a show of reason that Galileo now regretted that he had heedlessly lost the opportunity of attending the lectures. To make amends, he sought every opportunity to converse with the Copernicans, and remarking that they, like himself, had all once been Ptolemaists, and that from the doctrines of Copernicus no one had ever subsequently become a follower of the old philosophy, he resolved to examine the subject with the most serious attention. The result may be readily anticipated; the conversion was sudden and thorough, the old astronomy was abandoned, and the new convert became the great champion by whose ardor, and unconquerable zeal, the strongholds of antiquated systems were to be destroyed and a new and truthful one founded.

Thus far the career of Galileo in science had been successful and brilliant. He was rapidly rising in reputation and influence, when a fortunate accident revealed to the world, the application of a principle in optics fraught with consequences, which it is impossible to estimate. Galileo was informed that Jansen, of Holland, had contrived an instrument possessing the extraordinary property of causing distant objects, viewed through it, to appear as distinctly as when brought near to the eye. The extensive knowledge which Galileo possessed of optics, immediately gave him the command of the important principle on which the new instrument had been constructed. He saw at once the high value of such an instrument in his astronomical researches, and with his own hands commenced its construction.

After incredible pains, he finally succeeded in constructing a telescope, by whose aid, the power of the eye was increased thirty fold. It is impossible to conceive the intense interest with which the philosopher directed for the first time his wonderful tube to the inspection of the heavens. When we reflect that with the aid of this magical instrument the observer was about to sweep out through space, and to approach the moon, and planets, and stars, to within a distance only one-thirtieth of their actual distance; that their size was to increase thirty fold, and their distinctness in the same ratio, it is not surprising that these wonders should have excited the most extravagant enthusiasm.—Galileo commenced by an examination of the moon. Here he beheld, to his inexpressible delight, the varieties of her surface clearly defined, her deep cavities, her lofty mountains, her extensive plains, were distinctly revealed to his astonished vision. Having satisfied himself of the reality of these inequalities of the moon's surface, by watching the decreasing shadows of the mountains, as the sun rose higher and higher on the moon, he turned his telescope to an examination of the planets. These objects, which the human eye had never before beheld other than brilliant stars, now appeared round and clear and sharp, like the sun and the moon to unaided vision. On the 8th January, 1610, the telescope was for the first time directed to the examination of the planet Jupiter. Its disk was clearly visible, of a pure and silver white, crossed near the centre by a series of dark streaks or belts. Near the planet, Galileo remarked three bright stars which were invisible to the naked eye. He carelessly noted their position with reference to the planet, for he believed them to be fixed stars, and of no special interest, except to point out the change in Jupiter's place. On the following night, "induced," as he says, "by he knew not what cause," he again directed his attention to the same planet. The three bright stars of the preceding evening were still within the field of his telescope, but their positions with reference to each other were entirely changed, and such was the change, that the orbitual motion of Jupiter could in no way account for it. Astonished and perplexed, the eager astronomer awaits the coming of the following night to resolve this mysterious exhibition. Clouds disappoint his hopes, and he is obliged to curb his impatience.—The fourth night was fair, the examination was resumed, and again the bright attendants of Jupiter had changed—his suspicions were confirmed—he no longer hesitated, and pronounced these bright stars to be moons, revolving about the great planet as their centre of motion. A few nights perfected the discovery; the fourth satellite was detected, and this astounding discovery was announced to the world.

No revelation could have been more important or more opportune than that of the satellites of Jupiter. The advocates of the Copernican theory hailed it with intense delight; while the sturdy followers of Ptolemy stoutly maintained the utter absurdity of such pretended discoveries, and urged as a sort of unanswerable argument, that as there were but seven openings in the head—two ears, two eyes, two nostrils and the mouth, there could be in the heavens but seven planets. The more rational, however, saw the earth, by this discovery, robbed of its pretended dignity. It commanded the attendance of but one moon, while Jupiter received the homage of no less than four bright attendants. The delighted Copernicans saw in Jupiter as a central orb and in the orderly revolution of his satellites, a miniature of the sun and his planets, hung up in the heavens, and there placed to demonstrate to all coming generations, the truth of the new doctrines.

Another discovery soon followed, which it is said the sagacity of Copernicus foresaw would sooner or later be revealed to human vision. It had been urged by the Ptolemaists, that in case Venus revolved about the sun, as was asserted by Copernicus, and reflected to us the light of that luminary, then must she imitate exactly the phases of the moon; when on the side opposite to the sun, turning towards us her illuminated hemisphere she ought to appear round like the moon, while the crescent shape should appear on reaching the point in her revolution which placed her between the sun and the eye of the observer. As these changes were invisible to the naked eye, the objection was urged with a force which no argument could meet. Indeed it was unanswerable, and in case the telescope should fail to reveal these changes in Venus, the fate of the Copernican theory was forever sealed.

The position of Venus in her orbit was computed—the crescent phase due to that position determined—the telescope applied, and the eye was greeted with an exquisite miniature of the new moon. There was the planet, and there was the crescent shape long predicted by Copernicus, received by him and his followers as a matter of faith, now become a matter of sight. The doctrines of Copernicus thus received not only confirmation, but so far as Venus was concerned, a proof so positive that that no skepticism could resist. It is not my design to follow the discoveries of the Florentine philosopher among the planetary orbs. These will be resumed hereafter, when we come to examine more particularly the physical constitution of the planets. I have merely adverted to those discoveries, which became specially important in the discussions between the partisans of the old and new astronomy.

Admitting the doctrines of Copernicus, and uniting to them the great discoveries of Kepler, let us examine the condition of astronomical science, ascertain precisely the point the mind had reached, and the nature of the investigations which next demanded its attention. From the first of Kepler's laws, the figure of the planetary orbits became known, and the magnitude of the ellipse described by any planet was easily determined. By observing the greatest and least distances of any planet from the sun, the sum of these distances gave the longer axis of the orbit; and knowing this important line, and the focus, it became a simple matter to construct the entire orbit. The line joining the planet with the sun, while the planet occupied its shortest or perihelion distance, gave the position of the axis of the orbit in space, and its plane being determined by its inclination to that of the ecliptic, nothing remained to fix in space the figure, magnitude and position of the planetary orbits. The next point was to pursue and predict the movements of these revolving bodies. This was readily accomplished. A series of observations soon revealed the time occupied by any planet in performing one complete revolution in its elliptic orbit. Knowing thus the periodic time, and the position of the planet in its orbit at any given epoch, the second law of Kepler furnished the key to its future movements; its velocity in all parts of its orbit became known, and the mind swift and true followed the flying world in its rapid flight through space. It even went further; anticipated its changes, and predicted its positions, with a degree of certainty only limited by the accuracy with which the elements of its orbit had been determined.

The third of Kepler's laws, exhibiting the proportion between the periodic times and the mean distances of the planets from the sun, united all these isolated and wandering orbs into one great family. Their periods of revolution were readily determined by observation, and an accurate determination of the distance from the sun of a single planet in the group, gave at once the distance of all the remaining ones. The increased accuracy of the means of observation would render more perfect each successive measure of the earth's distance from the sun, and it seemed now that the mind might stop and rest from its arduous toil; that scarcely anything remained to be done. The solar system was conquered, and the fixed stars defied the utmost efforts of human power.

How widely does this view differ from the true one. In fact, the true investigation had not even commenced. A height had indeed been gained, from whence alone the true nature of the next great problem became visible, and standing upon this eminence the mind boldly propounds the following questions:—Why should the orbits of the planets and satellites be ellipses, rather than any other curve? What power compelled them to pursue their prescribed paths with undeviating accuracy? What cause produced their accelerated motion when coming round to those parts of their orbits nearer to the sun? What power held planet and satellite steady in their swift career, producing the most exquisite harmony of motion, and a uniformity of results as steady as the march of time?

Here I may be asked, do not such questions border on presumption? Are not such inquisitorial examinations touching on the domain of God's inscrutable providences, and would it not be wiser to stop and rest satisfied with the answer to all these questions, that God, who built the universe, governs and sustains it by his power and wisdom? Doubtless this answer is true, and in its truth man humbly finds his highest encouragement to attempt the resolution of the sublime questions already propounded for examination. Let us admit that the divine will produces all motion, speeds the earth in its rapid flight about the sun, guides the planets and their revolving moons, and poises the sun himself in empty space, as the great centre of life and light and heat to his attendant worlds. Is it not reasonable to believe that the will of the Omnipotent is exerted according to some uniform system, that this system is law, and that this law is within the reach of man? To encourage this view the simple laws of motion had been already revealed, and as these must exert a controlling influence in our future examinations, we proceed to unfold them.

First, then, it was discovered that if any body, situated in space and free to move, receive an impulse capable of giving it a velocity of ten feet in the first second of time, or any other velocity, it will move off in the direction of the impulse forever in a straight line, and with undiminished and unchanged velocity. The intensity of the primitive impulse determines the velocity of the body which receives it, and the one is precisely proportioned to the other. Again, in case a moving body while pursuing its flight receives an impulse in a direction different from its primitive one, its new direction and velocity will be determined by the direction and intensity of the new impulse, according to the principle discovered by Galileo, and already explained. Lastly, in every revolving body a disposition is generated to fly from the centre of rotation. The body seems urged by some invisible force from the centre, and if the velocity be sufficiently increased, no matter how strong the bond which unites it to the centre, it will, in the end, be severed, and the body, freed from its centre, darts away in a straight line tangent to its former circle of revolution. This power, which urges revolving bodies from the centre of motion, is called the *centrifugal* force, and is proportioned to the squares of the velocity of the revolving body. Hence a cord sufficiently strong to hold a heavy ball revolving round a fixed centre at the rate of fifty feet in a second, would require to have its strength increased four-fold, to hold the same ball, if its velocity should be doubled.

These simple laws, derived from a rigorous examination of those moving bodies, subject to man's closer scrutiny, extend their sway through the remotest regions of space. Are these laws necessary qualities of matter? Why should a body, darting away under the action of some impulsive force, pursue forever its undeviating direction, with undiminished velocity?—This effect cannot arise from any necessary property or equality of matter. The law might have been different—the direction of the moving body might have slowly varied—the velocity might have increased or decreased in any proportion, and yet the flying body, so far as we can understand, have retained all its physical qualities and properties. No—Divine wisdom has selected these simple and beautiful laws from among a multitude, either of which might have been chosen. Stretching forward, therefore, to the examination of the force by which the planets are retained in their orbits, was it not reasonable to expect, that some law might be found, governing the application of that mysterious power, and in some way proportioned to the mass of the moving body, and to the orbit which it described in wheeling around the sun. That they were held by some invisible power to their centre of motion, was manifest from the fact that the centrifugal force, generated by the rapidity of their revolution, would have hurled them away from the sun, if not opposed and counterpoised by some equivalent power lodged in the great centre of the planetary orbs. Here was an object worthy the highest ambition of the human mind.—No matter what might be the nature of this force, whether it should reside in the sun, or in the planet, or in both—whether it should prove to be a property of matter, or the mere uniform manifestation of the Omnipotent will; the discovery of its law of action would give to the mind the power of penetrating the darkest recesses of nature, and of rising to a knowledge of the profoundest secrets of the universe.

Such is the nature of the investigation propounded to the powerful intellect of Newton. This eminent philosopher, justly regarded as the most extraordinary genius that ever lived, neither originated the question which he undertook to discuss, nor divined the law of force which he proposed to demonstrate. When Kepler had closed the investigations which led to the discovery of his three great laws, his sagacity at once suggested to his mind the existence of some central force, by whose power the planetary movements were controlled. He had watched the moon circling around the earth, he had scrutinized the ocean tide, whose crested wave seemed to rise and follow the movements of the moon, until he boldly announced that some invisible bond, some inscrutable power, united the one to the other. He even reached the conclusion, that this unknown force resided in the moon—that by its power the waters were heaved from their beds, and caused to follow the moon and imitate its motions. Doubtless the solid earth itself felt this mysterious power, and swayed to its influence; but in consequence of the immobility of its particles, its effects had, thus far, escaped detection. Thus once started on the track, Kepler pursued the speculation. He attributed a similar power to the sun, and extended its controlling influence to the planets. He went yet farther, and conjectured that the law of this unknown force was such that it diminished as the squares of the distances at which it operated increased. That is, if the intensity of the power which it exerted on a planet where the distance was one hundred millions of miles from the sun, be counted as unity, removing the planet to double the distance, or to two hundred millions of miles, the sun's influence over it would be reduced to one-fourth of its former value.

With Kepler this wonderful conjecture always remained without proof. He had placed it on record, and succeeding philosophers had treated it with greater or less seriousness, according to the estimate which they placed upon the sagacity of its author. Even if Kepler, himself, had attempted the demonstration of this principle—the data were as yet wanting, which would have rendered its accomplishment possible. The period intervening between the time of Kepler and Newton had not been left unimproved. Descartes had revealed the law of centrifugal force, and by one of those extraordinary strokes of genius—occurring once in an age—had fastened the irresistible power of analysis upon geometry, which had given to the mind a force and rapidity in the investigation of the figure of curves and curvilinear motion, which had quadrupled its capacity. By repeated efforts, a more accurate knowledge had been obtained of the circumference and diameter of our earth, and through this the distance of the moon from the earth, in the successive points of its orbit had been approximated with still greater precision.

With these advantages, Newton gave the energies of his mind to the demonstration of that principle which had existed with Kepler as a mere conjecture.

Before proceeding to develop the train of reasoning pursued by the great English astronomer permit me first to prepare the way by a simple and perspicuous exhibition of the method employed in determining the diameter of the earth and the distance of the moon; two elements which figure conspicuously in the demonstration about to be made, and without a knowledge of which it would have been impossible to proceed. We commence with a determination of the diameter of the earth.

If an observer should start from any point on the surface of the earth in the northern hemisphere, and fixing his eye upon the north pole of the heavens, should travel directly towards that point, all the stars in the north would appear to rise higher above the horizon as he advanced in his journey. The star which occupied the point immediately above his head when he started, would appear gradually to decline towards the south. If it were possible to travel on the same great circle of the earth entirely around its circumference, the zenith star would appear to pursue an opposite route in the heavens, and would return to its primitive position only on the return of the observer to his point of starting. This, however, is not possible. What the observer can accomplish, is this. He may travel north until his zenith star shall appear to have moved south by one degree, or the three hundred and sixtieth part of the circumference of the heavens; then will he have passed over the three hundred and sixtieth part of the entire circumference of the earth; all these parts are of equal length,—he measures the one over which he has passed—multiplies its value by three hundred and sixty, and the result gives him the circumference of the earth, from which the diameter is readily deduced by the well-known proportion which exist between these lines. By this simple method, the diameter of the earth being determined, its radius is known, and we are prepared to explain the process by which the moon's distance may be found.

Let us locate, in imagination, two observers at distant points on the same

LAWS OF MOTION AND GRAVITATION. 69

great circle of the earth, each prepared to measure the angular distance at which the moon appears from the zenith point of each station; but the zenith of any place is the point in which the earth's radius prolonged reaches the heavens,—the angular distance of the moon from the zenith will exhibit precisely the inclination of the visual ray drawn to the moon's centre with the earth's radius drawn to the place of observation; the zenith distances being observed at each station, the observers knowing that part of the great circle of the earth by which their stations are separated, come together, compare observations, and construct a figure composed of four lines. Two of these are the radii of the earth drawn to the points of observation. These may be laid down under their proper angle,—drawing from their extremities two lines, forming with the radii, angles equal to the moon's measured zenith distances. These represent the visual rays drawn to the moon; they meet in a point which determines their length, and if the figure be constructed accurately, it will be found that either of these lines is about sixty times longer than the radius of the earth, or the moon's distance is about 240,000 miles.

We now return to the examination of the great question of a central force, and to the discovery of its law of action. Allow me in the out-set to explain, with extreme simplicity, the assumed law, whose truth or falsehood it was required to demonstrate. If any force resided in the sun which could resist the centrifugal force of the planets, or in any primary to resist the centrifugal force of the revolving satellite, it was conjectured that this force would decrease in proportion as the square of the distance increased. In other language, if the planets were arranged at the following distances from the sun, the forces exerted upon them would be represented by the second series, thus:

Distances, 1   2   3   4   5   6 &c.,
Forces,   1 $\frac{1}{4}$ $\frac{1}{9}$ $\frac{1}{16}$ $\frac{1}{25}$ $\frac{1}{36}$ &c.

The measure of the intensity of any force of attraction situated at the centre of the earth or sun, is accurately represented by the velocity it is capable of imparting to a falling body in any unit of time. Experiment shows that that power which causes a heavy body to fall to the earth's surface, is capable of impressing upon it a motion of about 16 feet in the first second of time after its fall commences. In case the force diminishes, as we remove the falling body farther from the centre of attraction, the law of diminution would manifest itself in the diminished amount of motion communicated to the falling body.

Now if Newton could have carried a heavy body upward, above the earth, until he should gain a height above its surface of four thousand miles, he would then be twice as far from the centre as when at the surface of the earth. Dropping the heavy body, and measuring accurately the distance through which it passes in the first second of time, in case he finds this to be *one-fourth* of 16 feet, the distance fallen through by the same body in the same time, at the distance of 4000 miles from the earth's centre, the result would have confirmed the law which conjecture had assigned as the law of nature. Could he have mounted one unit higher, gaining an altitude of 8000 miles above the earth's surface, or *three* units from the centre, here repeating his experiment, in case the space passed through by the falling body is now *one-ninth* of 16 feet, it would yet farther confirm the truth of the conjectured law. Thus, could he have increased his altitude by one unit or radius of the earth after another, repeating his experiment as each new unit was added to his elevation, finding in every instance the law of diminution fulfilled by the falling body, all doubt as to the truth of the law would have been removed, and its foundation in nature would have justly flowed from such a series of experiments.

Here, then, is precisely what must be accomplished to demonstrate the assumed law of gravitation. But since these altitudes of 4000 and 8000 miles could not be reached, might not some change in the distance passed over by a heavy falling body, be noticed and measured, if removed from a valley to the top of the highest mountain? Alas! the increased distance from the centre of the earth, gained by ascending the loftiest mountain on its surface, is almost inappreciable, when compared with the entire distance, four thousand miles. Even if the mountain were ten miles high, the two elevations at which the experiment might then be performed, would be 4000 miles and 4010 miles, and the diminished velocity would not be appreciable, even with the most delicate tests, much less could it be relied on to demonstrate the truth or falsehood of a great principle. Here, then, the mind was brought to a full stop; and for a long time it seemed impossible that the philosopher should conquer the difficulties which rose up in his path, and defied his further advance. Finding it impossible to perform any satisfactory experiment on the earth's surface, the daring mind of Newton conceived the idea of employing the moon itself as the falling body, and of testing the truth of his great theory by its fall towards the earth. But could he reach out his hand, grasp the revolving moon, stop it in its orbit, drop it to the earth, and measure its descent in the first second of time? No—this was impossible. The moon could not be arrested in its career; but is this necessary? Is not the moon, in one sense, constantly falling towards the earth? Newton asserted this to be true, and thus did he prove it.

Stand upon the earth, and stretching outward into space 240,000 miles, there let the moon be located, poised and fixed in space, on a point of its present orbit. There let us suppose it to receive an impulse in a direction perpendicular to the line which joins it with the earth. By the first law of motion, being free to move, it will sweep off in a straight line, tangent to its present orbit, and will pass over a space in the first second of time, proportioned to the intensity of the impulse received. Mark that space, and bring the moon back to its primitive position. Now drop it towards the earth, and as it descends freely under the earth's attraction, mark the space through which it falls in the first second of time. This being known, bring back the moon once again to its starting-point. Now combine the impulsive force first given with that power which caused the moon, when left free to move, to fall to the earth. Let them both act at once: the impulse is given, the moon darts off in a straight line, but is instantly seized by the earth's attraction, which drags it from its rectilineal path, and the two contending forces, ever struggling, neither conquering, exercise a divided empire over the moon; onward she moves, obedient to the impulsive force, bent to her orbit by the action of the earth's attraction. Now the amount by which it is deflected in one second of time, from the straight line it would have pursued, is the *amount precisely, by which it falls to the earth.*

If thus far I have been successful, what remains can readily be accomplished. Newton easily computed, from the known velocity of the moon in its orbit, and from the radius of that orbit, the space through which the moon actually fell towards the earth in one second of time. He next computed the space through which a heavy body would fall towards the earth's surface, if removed from the earth to a distance equal to that of the moon. Now in case these two quantities should prove to be exactly equal, the truth of the demonstration would be complete; the moon did fall through the space required by the assumed law, and in this event the law must be the law of nature. For seventeen long years did this incomparable philosopher, rivaling the example of the immortal Kepler, toil on in this most difficult enterprise. He finally reaches the result; the two quantities are found and compared, but alas! the computed distance through which the moon must fall, in case the law of gravitation were true, differed from the observed distance through which it actually fell, by a sixth part of its value. Any less scrupulous, any less philosophic mind, would have been content with this near approximation, and would have announced the discovery to the world. Not so with Newton. Nothing short of the most rigorous accuracy could satisfy his conscientious regard for truth. His manuscripts are laid aside, and the pursuit for the present abandoned.

Months roll by. Occasionally he returns to his computations, runs over the figures, hoping to detect some numerical error; but all is right, and he turns away. At length, while attending a meeting of the Royal Society in London, he learns that Picard had just closed a more accurate measurement of the diameter of the earth. This was one of the important quantities which entered into his investigation. He returns home,—and with impatient curiosity spreads before him his old computations—the new value of the earth's diameter is substituted—he dashes onward through the maze of figures—he sees them shaping their value towards the long sought result—the excitement was more than even his great mind could bear—he resigns to a friend—the work is completed, the results compared—they are exactly equal! The victory is won,—he had seized the golden key which unlocks the mysteries of the universe, and he held it with a giant's grasp!

There never can come another such moment as the one we have described, in the history of any mortal. There are no such conquests remaining to be made. Standing upon the giddy height he had gained, Newton's piercing gaze swept forward through coming centuries, and saw the stream of discovery flowing from his newly discovered law, slowly increasing, spreading on the right hand and on the left, growing broader, and deeper, and stronger, encircling in its flow planet after planet, sun after sun, system after system, until the universe of matter was encompassed in its mighty movement. He could not live to accomplish but a small portion of this great work.—Rapidly did he extend his theory of gravitation to the planets and their satellites. Each accorded perfectly with the law, and rising as the inquiry was pursued, he at length announced this grand prevailing law:

Every particle of matter in the universe attracts every other particle of matter with a force or power directly proportioned to the quantity of matter in each, and decreasing as the squares of the distances which separate the particles increase.

Having reached this wonderful generalization, Newton now propounded this important inquiry. "To determine the nature of the curve which a body would describe in its revolution about a fixed centre, to which it was attracted by a force proportional to the mass of the attracting body, and decreasing with the distance according to the law of gravitation."

His profound knowledge of the higher mathematics which he had greatly improved, gave to him astonishing facilities for the resolution of this great problem. He hoped and believed that when the expression should be reached, which would reveal the nature of the curve sought, that it would be the mathematical language descriptive of the properties of the ellipse. This was the curve in which Kepler had demonstrated that the planets revolved, and a confirmation of the law of gravitation required that the ellipse should be the curve described by the revolving body, on the conditions announced in the problem.

There happens to be a remarkable class of curves, discovered by the Greek mathematicians, called the *conic sections;* thus named, because they can all be formed by cutting a cone in certain directions. The figure of a cone with a circle for its base, and converging to a point, is familiar to all. Cut this cone perpendicular to its axis, remove the part cut, and the line on the surface round the cone will be found to be a *circle.* Cut it again, oblique to the axis, then the line of division of the two parts will be an *ellipse.* Cut again so that the knife may pass downward parallel to the slope of the cone, and in this case your section is a *parabola.* Make a last cut parallel to the axis of the cone, and the curve now obtained is the *hyperobla*.

When Newton reached the algebraic expression which, when interpreted, would reveal the properties of the curve sought and which he had hoped would prove to be the ellipse—he was surprised to find that it did not look familiar to his eye. He examined it closely—it was not the equation of the ellipse, and yet it resembled it in some particulars. What was his astonishment to find, on a complete examination, that the mathematical expression, which he had reached, expressing the nature of the curve described by the revolving body, was the general algebraic expression embracing *all the conic sections*. Here is a most wonderful revelation. Is it possible that under the law of gravitation, the heavenly bodies may revolve in any or either of these curves? Observation responds to the inquiry. The planets were found to revolve in ellipses; the satellites of Jupiter in circles; and those strange, anomalous, outlawed bodies, the *comets*, whose motions hitherto had defied all investigation, take their place in the new and now perfect system, sweeping round the sun in *parabolic* and *hyperbolic* orbits.

Thus were these four beautiful curves, having a common origin, possessed of certain common properties, yet diverse in character, mingling in close proximity, and gliding imperceptibly into each other, suddenly transferred to the heavens, to become the orbits of countless worlds. For nearly twenty centuries, they had been the objects of curious speculation to the mathematician; henceforward they were to be given up to the hands of the astronomer, the powerful instrument of his future conquests among the planetary and cometary worlds.

The three great laws of Kepler, to which he had risen with such incredible toil and labor, were now found to flow as simple consequences of the law of gravitation. It is impossible to convey the slightest idea, in discussions so devoid of mathematics, of the incredible change which had thus suddenly been wrought in the mode of investigation. I never have closed Newton's investigation, by which he deduces the nature of the curve, described by a body revolving around a fixed centre, under the law of gravitation, bearing with it consequences so simple yet so wonderful, without feelings of the most intense admiration. I can convey no adequate idea of the difference of the methods employed by Kepler and Newton, in reaching the three laws of planetary motion. I see Kepler in the condition of one on whom the fates have fixed the task of rolling a huge stone up some rugged mountain side, to its destined level, within a few feet of the summit. He toils on manfully, heaving and struggling, day and night, in storm and in darkness, never quitting his hold, lest he may lose what he has gained. If the ascent be too steep and rocky, he diverges to the right, then to the left, winding his heavy way zigzag up the mountain side.—Years glide by—he grows gray in his toil, but he never falters—onward and upward he still heaves the heavy weight—his goal is in sight, he renews his efforts, the last struggle is over—he has finished his task—the goal is won.

Such was Kepler's method of reaching his laws. Now for Newton's. He stands, not at the base of the mountain, with its long, ascending rocky sides, but on the top. He starts his heavy stone, it rolls of itself over, slowly over, and once again, and falls quietly to its place. Let me not be misunderstood in this strange comparison, as detracting in the smallest degree from the just fame that is due to Kepler. But for his sublime discoveries, Newton could never have reached the mountain summit, on which he so proudly stood. Standing there, he never forgot by whose assistance he had reached the lofty point, and ever recognized, in the most public manner, his deep indebtedness to the immortal Kepler.

A few words with reference to the rigorous application of Kepler's laws in nature, will close this discussion. The first law, announcing the revolution of the planets in elliptic orbits, was now made general, and recognized the revolution of the heavenly bodies in *conic sections:* the circle, ellipse, parabola and hyperbola.

The second law, fixing the equality of the spaces passed in equal times, by the lines joining the planets to the sun as these were carried round in their elliptic orbits, now became applicable to all bodies revolving about a fixed centre, in any curve, and according to any law.

The third law, recognizing the proportion between the squares of the periodic times and the cubes of the mean distances of the planets, was extended to the satellites, and to the comets; modified slightly in the case of the larger planets, by taking into account their masses or quantities of matter.

Here we close the era of research by observation. The mind has gained its last grand object. The era of physical astronomy dawns; new and wonderful scenes open, and to the contemplation of these we shall soon invite your attention.

LECTURE V.

UNIVERSAL GRAVITATION APPLIED TO THE EXPLANATION OF THE PHENOMENA OF THE SOLAR SYSTEM.

THE progress of the mind, in its efforts to reach a satisfactory explanation of the movements of the heavenly bodies, previous to the discovery of universal gravitation, had been made independent of any guiding law. The mind had been feeling its way slowly and laboriously, guiding its direction by attentively watching the celestial phenomena, and relying for its success exclusively on the accuracy and number of its observations. Each discovery made was isolated, and although it prepared the way for the succeeding ones, it did not in any sense involve them as necessary consequences. By the discovery of the great law of universal gravitation, a perfect and entire revolution had been made in the science of astronomy. A new department was now added, which, previous to the knowledge of this law, could have no existence. In this branch of astronomy, the process of investigation being inverted, the mind descends from one great law to an examination of its consequences, tracing these in their modified and diversified influences to their final limits. Observation is now employed to verify discovery and not as the basis on which, and without which, discovery cannot be made.

The era of physical astronomy is, therefore, the great era in the history of the science. It involves the resolution of the most wonderful problems—it calls into use the most refined and powerful mathematical analysis, and demands the application of the most ingenious and delicate instrument in seeking for the data by means of which its theory may be rendered practically applicable to the problems of nature. The mechanical philosopher in his closet may construct his imaginary system. In its centre he locates a sun, containing a certain mass of matter.—At any convenient distance from this sun he locates a planet, whose weight he assumes. To this planet he gives an impulse, whose intensity and direction are assumed. The moment these data are fixed, and the impulse given to the imaginary planet—no matter in what kind of an orbit it may dart away, whether circular, elliptical, parbolic or hyperbolic—the laws of motion and gravitation asserting their empire, the planet is followed by the mathematician, with a certainty and accuracy defying all escape. He assigns its orbit in the heavens—the velocity of its movement—the period of its revolution. In short, in a single line, he writes out its history with perfect accuracy for a million years.

If, now, to this simple system of a great central sun and one solitary planet, the physical astronomer add a third body, a moon, to the planet, he assumes its weight, the intensity and direction of the impulsive force starting it in its career, and now his system becomes more complex. Strike the sun out of existence, fix the planet, and the process of binding the satellite in mathematical fetters is precisely similar to that by which the movements of the planet were prescribed around the sun before the existence of the satellite. But now with these three bodies the train of investigation becomes more intricate and involved. While the planet alone circulated around the sun, such is the undeviating accuracy with which it will forever pursue its path around the sun, that if it were possible to hang up in space along its route golden rings whose diameter would just permit the flying planet to pass, millions of revolutions will never mark the slightest change. The rings once passed and then fixed, will mark forever the pathway of the solitary planet. But the moment a moon is given to this flying world, in that instant its motion is changed—it is swayed from its original fixed orbit—it no longer passes through the golden rings, and although the physical astronomer may write out in his analytic symbols the future history of his planet and moon, these expressions are no longer marked with the simplicity which obtained in those which recorded the history of the single planet. While solitary, all changes were effected by the planet in one single revolution, and these were repeated in the same precise order in each successive revolution. Now, with the satellite added, there are changes introduced running through many revolutions, and requiring for their complete compensation vast periods of time. Indeed, the inquiry arises, whether this system of a central sun, with a planet and its satellite revolving about it, can be so constituted that the changes which the planet and its moon mutually produce on each other's movements may not go on constantly accumulating in the same direction until all features of the original orbits of both may be destroyed, both worlds being finally precipitated on the sun, or driven farther and farther from this luminary, until they are lost in infinite space. This inquiry, in a more extended form, will be examined hereafter. We proceed to build up our imaginary system. Thus far we have regarded our planet and its satellite as mere material heavy points. In case we give to them magnitude and rotation on an axis, the velocity of rotation will determine the figure of the planet and of its satellite. These figures will deviate from the exact spherical form, and this change of figure will sensibly affect the stability of the axis of rotation, and will introduce a series of subordinate movements, each of which must become the subject of research ; and to write out the future history of the system these minute and concealed changes must likewise receive their mathematical expression.

Having thus thoroughly mastered all the phenomena of this system of three bodies, the astronomer now adds another planet, whose mass is assumed, together with the direction and intensity of its primitive impulse. Its orbit is now computed, subject to the greatly predominant influence of the sun, but sensibly affected by the quantity of matter in the old planet and its satellite, which prevents it from forming a fixed and unchangeable orbit in space. Again he is obliged to return to an examination of his first planet and its moon, for these again break away from their previous routes, and in consequence of the action of the second planet, assume new orbits, and are subjected to periodical fluctuations, which demand critical examination, and without a knowledge of which no truthful history of the planetary system can be written. To the second planet let us now add several satellites, each of which has its mass assigned, and the direction and intensity of the impulsive force by which they are projected in their orbits. Here, then, is a subordinate system demanding a complete examination. The satellites mutually affect each other's motions, and each is subjected to the influence of the primitive planet and its moon. Again does the physical astronomer review his entire investigation. The addition of these satellites to his second planet has introduced changes in all the previous bodies of the system, which must now be computed, to keep up with the growing complexity. This task is at last accomplished. All the changes are accurately represented. Analysis has mastered the system, and the history of its changes are written out for hundreds and thousands of years.

A third planet, with its satellite, is now added. This new subordinate system is discussed, and its operation on all the previous planets and satellites computed, and after incredible pains, the astronomer once more masters the entire group, and follows them all with unerring precision, through cycles of changes comprehending thousands or even millions of years.

Thus does the difficulty of grasping the system increase in a high ratio by the addition of every new planet and satellite, till finally the last one is placed in its orbit, and the system is complete, so far as planets and satellites are concerned. Through this complicated system now cause thousands of comets to move in eccentric orbits, coming in from every quarter of the heavens, plunging downwards towards the sun, sweeping with incredible velocity around this central luminary, and receding into space to vast distances, either to be lost forever, or to return after long periods to revisit our system. These wandering bodies must be traced and tracked, their orbits fixed, their periods determined, their influence on the planets and satellites, and that exerted by these on the comets, must be computed and determined; then, and not till then, does the physical astronomer reach to a full knowledge of this now almost infinitely complex system.

In this imaginary problem it will be observed that certain quantities were invariably assumed before the discussion could proceed. The mass of the sun—the mass of each planet and satellite—the intensity and direction of the primitive impulse given to each planet and satellite—these quantities are supposed to be known. If, now, the astronomer has actually accomplished the resolution of the imaginary problem, and has obtained analytic expressions which write out and reveal the future history of his assumed planets and satellites, as they revolve around his assumed sun, if in these expressions he should substitute the actual quantities which exist in the solar system for those assumed, his expressions would then give the history of the solar system for coming ages, and by reverse action would reveal its past history with equal certainty.

Before we can, therefore, bring the power of analysis to bear on the resolution of the grand problem of nature, we must interrogate the heavens, and obtain the absolute weight of our sun, of each planet, and of every satellite. Next we require the intensity and direction of the impulsive force which projected each planet and satellite in its orbit, and which would have fixed forever the magnitude and position of that orbit, in case no disturbing causes had operated to modify the action of the primitive impulse.

Having thus attempted to exhibit, at a single view, the general outlines of the great problem of the solar system, we propose now to return to the examination of a system composed of three bodies; and to fix our ideas, we assume the sun, earth, and moon. In case the earth existed alone, the elliptic orbit described in its first revolution around the sun would remain unchanged forever, and having pursued it, and marked its changes of velocity in the different parts of its orbit for a single revolution, this would be repeated for millions of years. But let us now give to the earth its satellite, the moon, and setting out from its perihelion, or nearest distance from the sun, let us endeavor to follow these two bodies as they sweep together through space, and mark particularly the effect produced on the moon's orbit by the disturbing influence of the sun. To give to the problem greater simplicity, let us conceive the plane of the moon's orbit to coincide with the earth's. The law of gravitation which gives to every attracting body a power over the attracted one, gravity increasing as the distance decreases, it will be perceived that when the earth and moon are nearest to the sun, whatever influence the sun possesses to embarrass or disturb the motions of the moon about the earth, will here be exercised with the greatest effect. But since the sun is exterior to the moon's orbit, its tendency will be to draw the moon away from the earth, and cause her to describe around her primary a larger orbit, in a longer period of revolution than would have been employed in case no sun existed, and the moon was given up to the exclusive control of the earth.

et on its annual journey, as it recedes from the sun in
rest to its most remote distance, or from perihelion to
is gradually removed from the disturbing influence of
ted more exclusively to the earth's attraction; its dis-
grows less, and the periodic time becomes shorter.
tinue in the same order until the earth reaches its
he moon's orbit is a minimum, and its motion is
g from the aphelion to the perihelion, the earth is con-
the sun, and as the sun's influence on the moon in-
nce diminishes, its orbit will now expand by slow
riodic time will diminish until on reaching the peri-
figure of the earth's orbit remains unchanged, the
e will be restored to its primitive value, and all the
m the elliptic figure of the earth's orbit will have
l.

directed our attention exclusively to the changes in
moon and its periodic time. But the moon's orbit is
the earth's, and it is manifest that the sun's influence
re not only the magnitude of this orbit, but will in like
hange in the position of the moon's perigee, or nearest
m the earth. If the earth were stationary, and the
nd it, passing between it and the sun, and then com-
be beyond the earth with respect to this luminary,
orbit would be sensibly affected by the sun's attrac-
exerted itself during one revolution of the moon, all
repeated in the same order during the next revolu-
e positions of the sun and earth remaining the same,
e finally to have a fixed orbit, and its principal lines
r change. But this is not the case of nature.—The
in its orbit, and bearing with it its revolving satellite,
n has completed a revolution, the sun and earth have
eir relative positions, and the moon cannot reach its
listance from the earth, at the same point as in the
l.

xamination of this problem, it is found that the tend-
moon to reach its perigee earlier than it would do if
n this way the perigee of a fixed orbit appears to ad-
ming moon, and in the end to continue advancing
lves entirely round in a period which observation de-
t nine years.

tion to enter into a detailed examination of all the
n the sun's disturbing power on the moon's motions;
pt to exhibit all the effects produced by the moon on
uld require a train of investigation too elaborate and

intricate to comport with my present purposes. My object is simply to show that changes must arise from the mutual and reciprocal action of these three bodies, which the theory of gravitation must explain, and the telescope point out, before it be possible to obtain a perfect knowledge of these bodies.

The exact estimation of these changes can never be made until we shall have learned the relative masses of matter contained in the sun, earth, and moon. In other language, we must know how many moons it would require to weigh as much as the earth, and how many earths would form a weight equal to that of the sun.

But is it possible that man, situated upon our planet, 237,000 miles from the moon, and 95,000,000 of miles from the sun, can actually weigh these worlds against each other, and determine their relative masses of matter? Even this has been accomplished, and I shall now proceed to explain how the earth may be weighed against the sun. Dropping a heavy body at the earth's surface, the velocity impressed on it in the first second of time will measure the weight of the earth in one sense. If it were possible to take the same body to the sun, drop it, and measure the velocity acquired by the falling body in the first second of time, the relative distances passed through at the sun and at the earth by the same body in the same time, would show exactly the relative weight of the sun and earth, for their capacity to communicate velocity are exactly proportioned to their masses. Now, although this experiment cannot be performed in the exact terms announced, yet as we have already shown, the moon is constantly dropping towards the earth, and the earth is as constantly dropping towards the sun. Now in case we measure the amount by which the moon is deflected from a straight line in one second of time, this measures the intensity of the earth's power. But the amount by which the earth is deflected from a right line by the central power of the sun in one second, is easily measured from a knowledge of its period and the magnitude of its orbit. Executing these calculations, it is found that the sun's effect on the earth is rather more than twice as great as the earth's effect on the moon, and in case these effects were produced at equal distances, then would the sun be shown to contain rather more than twice as much matter as is found in the earth. But the sun produces its effect at a distance 400 times greater than that at which the earth acts on the moon; hence, as the force diminishes as the square of the distance increases, a sun acting at twice the distance at which the earth acts, must be four times heavier to produce an equal effect; at three times the distance, it must be nine times heavier, and at four times the distance, sixteen times heavier; —at 400 times the distance, 160,000 times heavier than the earth. Thus do we find that in case the sun's action on the earth were exactly equal to the earth's action on the moon, in consequence of the great distance at which it operates, its weight would be equal to that of 160,000 earths.

But its actual effect is rather more than double that of the earth on the moon, and hence we find it contains rather more than double 160,000 earths, or exactly 354,936 times the quantity of matter contained in the earth.

This enormous mass of the sun is confirmed by an examination of its actual dimensions. An object with an apparent diameter equal to that of the sun and at a distance of 95,000,000 of miles, must have a real diameter of 883,000 miles, a quantity so great that if the sun's centre were placed at the earth's centre, its vast circumference would give ample room for the moon to circulate within its surface, leaving as great a space between the moon's orbit and the sun's surface as now exists between the moon and earth.

It is this immense magnitude of the sun, when compared with the planets and their satellites, which renders the orbits of the planets comparatively unalterable. It is true that these bodies mutually affect each other, but these effects are comparatively slight, and astronomers regard them as *perturbations*, or mere disturbances of the original elliptic motion. Hence we find the magnitude and position of the earth's elliptic orbit remain without any very sensible variation for two or three revolutions; but the slight disturbance experienced at each revolution, constantly accumulating in the same direction in a long series of years, occasions changes that cannot be lost sight of, and which, by a reflex influence, become in some instances exceedingly important in their practical applications.

As it will be impossible to treat fully the complex subject of perturbation, I will call your attention to a few points about which cluster peculiar interests, in consequence of their great difficulty, and the almost infinite reach of analysis displayed in their successful examination.

I have already explained how it is that the disturbing influence of the sun occasions a constant fluctuation in the periodic time of the moon, accelerating it as the earth moves from perihelion to aphelion, and again retarding it from aphelion to perihelion. If, now, we take a large number of revolutions of the moon, say a thousand, add them all up, and divide by one thousand, we obtain a mean period of revolution, which, in case the earth's orbit remains invariable will never change, but will be constantly the same for thousands of years. By such an examination during the last century, the mean motion of the moon was obtained with great precision. But on a comparison of eclipses recorded by the Babylonians with each other, it was discovered that the moon in those early ages required a longer time to perform her mean revolution than in modern times. A like comparison of the Babylonian eclipses with those recorded in the middle ages by the Arabian astronomers, confirmed this wonderful discovery, which was yet farther substantiated by comparing the Arabian eclipses with those observed in modern times. It thus became manifest that, to all appearance at least, the moon's mean motion was growing swifter and swifter from century to century; that it was approaching closer and still closer to the earth, and if no limit to this change was ever to be fixed, sooner or later the final catastrophe must come, and the moon be precipitated on the body of the earth, and the system be destroyed.

An effort was made to account for this acceleration of the moon, on the theory of gravitation; but for a long time there seemed to be no possibility of rendering a satisfactory explanation of the phenomena, far less of prescribing the limits which should circumscribe the changes. Some, to escape from the difficulty, rejected entirely the ancient eclipses, and boldly cut the knot, by pronouncing the acceleration as impossible, and without any foundation in fact.—Others admitted the fact, but finding it impossible to account for it on the hypothesis of gravitation, conceived the idea that the moon was moving in some ethereal fluid capable of resisting its motion, and producing a diminution in its periodic time of revolution. That *acceleration* should be the effect of *resistance*, may seem to some very strange, but a little reflection will render the subject clear. In case the moon's orbitual motion is resisted, then the centrifugal force, which depends on the velocity, becomes diminished, and the central power of the earth draws the moon closer to itself, decreases the magnitude of its orbit, and in like manner reduces the time of accomplishing one revolution about the earth.

Finding no better solution of the mystery, and being obliged to acknowledge the fact that the mean motion of the moon was becoming swifter and swifter, from the action of a resisting medium, there was no escape from the final consequences; and it was by some believed that the elements of decay existed, that the doom of the system was fixed, and although thousands, possibly millions, of years might roll away before the fatal day, yet it must come, slowly, but surely as the march of time. Such was the condition of the problem when Laplace gave the powers of his giant intellect to the resolution of this mysterious subject. The consequences involved gave to it an unspeakable interest, and the world waited with keen anxiety to learn the result of the investigations of this great geometer. Long and difficult was the struggle—slow and laborious the task of devising and tracing out the secret causes of this inscrutable phenomenon. The planets are weighed and poised against the earth, their effects computed on its orbit, the final result of these effects determined, and the reflex influence on the moon's motion computed with the most extraordinary precision. Under the searching examination of Laplace's potent analysis, nature is conquered, the mystery is resolved, the law of gravitation is vindicated— the system is stable, and shall endure through periods whose limits God alone, and not man, shall prescribe.

Follow me in a simple explanation of this most remarkable discovery. It has already been stated, that in case the earth's orbit could remain unchanged, that the mean period of the moon, as derived from a thousand of its revolutions, would be constant, and would endure without the slightest change for millions of years. But this permanency of the earth's orbit does not exists.—Laplace discovered that under the joint action of all the planets, the figure of the earth's orbit was slowly changing; that while its longer axis remained invariable, that its shape was gradually becoming more and more nearly circular. At the end of a vast period, its ellipticity would be destroyed, and the earth would sweep around the sun in an orbit precisely circular. Attaining this limit, a reversed action commences—the elliptic form is resumed by slow degrees—the eccentricity increases from age to age—until, at the end of millions of years, a second limit is reached. The motion is again reversed—the orbit again opens out, approaches its circular form, and thus vibrates backwards and forwards in millions of years, like some mighty pendulum beating the slowly ebbing seconds of eternity!

But do you demand how this change in the figure of the earth's orbit can effect the moon's mean motion? The explanation is easy. Were it possible to seize the earth and hurl it to an infinite distance from the sun, its satellite, now released from the disturbing influence of this great central mass, would yield itself up implicitly to the earth's control. It would be drawn closer to its centre of motion, and its orbit being thus diminished, its periodic time would be shorter, or its motion would be accelerated or made swifter than it now is. This is an exaggerated hypothesis, to render more clear the effect produced by removing the earth farther from the sun. Now the change from the elliptical to the circular form, which has been progressing for thousands of years, in the earth's orbit, is, so far as it goes, carrying the earth at each revolution a little farther from the sun, releasing in this way the moon, by slow degrees, from the disturbing influence of that body; giving to the earth a more exclusive control over the movements of its satellite, and thus increasing the velocity of the moon in its orbit from age to age. But will this acceleration ever reach a limit? Never, until the earth's orbit becomes an exact circle, at the end of millions of years. Then, indeed, does the process change. At every succeeding revolution of the earth in its orbit, its ellipticity returns—its distance from the sun diminishes—the moon is again subjected more and more to the action of the sun, is drawn farther and farther from the earth, and its periodic time slowly increases. Thus is acceleration changed into retardation, and at the end of one of these mighty cycles, consisting of millions of years, an exact compensation is effected, and the moon's motion having gone through all its changes, once more resumes its original value.

I can never contemplate this wonderful revolution without feelings of profound admiration. Such is the extreme slowness of this change in the moon's mean motion, that in the period of three thousand years she has got only four of her diameters in advance of the position she would have occupied in case no change whatever had been going on. Here, then, is a cycle of changes extending backward to its least limit millions of years, and extending forward to its greatest limit tens of millions of years, detected and measured by man, the existence of whose race on our globe has scarcely been an infinitesimal portion of the vast period required for the full accomplishment of this entire series of changes.

May it not, then, be truly said, that man is in some sense immortal, even here on earth. What is time to him, who embraces changes in swiftly revolving worlds, requiring countless ages for their completion, within the limits, of an expression so condensed that it may be written in a single line? Does he not live in the past and in the future, as absolutely as in the present? Indeed, the present is nothing—it is the past and future which make up existence.

In the example of the moon's acceleration just explained, we must not fail to notice a most remarkable fact. It is this:—The slow change in the figure of the earth's orbit, occasioned by the joint action of all the planets, and upon which depends the acceleration of the moon's mean motion, is so disguised, that but for its reflex influence on the moon, the probability is it would have escaped detection for thousands of years. The direct effect is almost insensible, but being indirectly propagated to the moon, it is displayed in a greatly exaggerated manner—is in this way detected, and finally, after incredible pains, traced to its origin, and demonstrates in the most beautiful manner the prevalence of the great law of universal gravitation.

Since the general adoption of this law, the human mind has been, in not a few instances, disposed to abandon its universality, and seek for a solution of some intricate problem, by which it was perplexed, in some change or modification of the law; but in no instance has the effort to fly from the law been successful. No matter how long and intricate the examination, how far the mind might be carried from this great law, in the end it must come back and acknowledge its universal empire over our entire system.

It has already been remarked, that one of the effects of the sun's disturbing influence exerted on the moon, was to occasion a change in the position of its perigee, causing it to complete an entire revolution in the heavens in about nine years. The theory of gravitation gave a very satisfactory account of this phenomenon generally; but when Sir Isaac Newton undertook the theoretic computation of the rapidity with which the moon's perigee should move, he found, to his astonishment, that no more than one half of the observed motion of the perigee was obtained from theory. In other language, in case the law of gravitation be true, Newton found that the moon's perigee ought to require eighteen years to perform its revolution in the heavens, while observation showed that the revolution was actually performed in one half of this period. This great philosopher exhausted all his skill and power in the vain effort to overcome this difficulty. He died, leaving the problem unresolved, bequeathing it to his successors, as a research worthy of their utmost efforts.

Astronomers did not fail to recognize the high claims of this investigation. Gravitation was once more endangered. The most elaborate computations were made, and the results obtained by Newton were so invariably verified by each successive computer, that it seemed utterly impossible to avoid the conclusion;—they were absolutely accurate, and that the theory of gravitation must be modified in its application to this peculiar phenomenon. At length the problem was taken up by the distinguished astronomer Clairaut. After repeating, in the most accurate manner, the extensive computations of his predecessors, reaching invariably the same results, he finally abandoned the law of gravitation in despair, pronounced it incapable of explaining the phenomenon, and undertook to frame a theory which should be in accordance with the facts.

This startling declaration of Clairaut excited the greatest interest. An abandonment of the theory of gravitation was nothing less than returning once more to the original chaos which had reigned in the planetary worlds, and of commencing again the resolution of the great problem which it had long been hoped was entirely within the grasp of the human intellect. In this dilemma, when the physical astronomer had abandoned the law of gravitation in despair, and the legitimate defenders of the theory were mute, an advocate arose where one was least to be expected. Buffon, the eminent naturalist and metaphysician, boldly attacked the new theory of Clairaut, pronounced it impossible, and defended the law of gravitation by a train of general reasoning, which the astronomer felt almost disposed to treat with ridicule. What should a naturalist know of such matters? was rather contemptuously asked by the astronomer. It is true he knew but little, yet his attack on Clairaut had the effect to induce the now irritated astronomer to return to his computations, with a view to overwhelm his adversary. He now determined to rest satisfied with nothing short of absolute perfection.—A certain series which had been reached by every computer, and the value of whose terms had been regarded as decreasing by a certain law, until they finally became inappreciable, from their extreme minuteness, and therefore might, without sensible error, be rejected, was found, on a more careful examination, to undergo a most remarkable change in its character. It was true that the value of its terms did decrease till they became exceedingly small; but so far from becoming absolutely nothing, on reaching a certain value, the decrease became changed into increase—the sum of the series expressing the velocity of the moon's perigee was in this way actually doubled, and Clairaut found, to his inexpressible astonishment, that the investigation which had been commenced with the intention of forever destroying the universality of the law of gravitation, resulted in his own defeat, and in the perfect and triumphant establishment of this great law.

Thus far, in our examinations of the moon and earth, we have regarded their orbits as lying in the same plane, an hypothesis which greatly simplifies the complexity of their motions. This, however, is not the case of nature. The moon revolves in an orbit whose plane is inclined under an angle of about four degrees to the plane of the ecliptic. During half of its journey, it lies above the plane of the earth's orbit, while the remaining part of its route is performed below the ecliptic. Thus does the moon, at each revolution, pass through the ecliptic at two points, called the *nodes*, which points, being joined by a straight line, gives us the intersection of the plane of the moon's orbit with that of the earth. This line of intersection, called the line of the nodes, but for the disturbing influence of external causes, would remain fixed in the heavens. But we know it to be constantly fluctuating, and in the end performing an entire revolution. The exact amount of this change has been made the subject of accurate examination, and the law of its movement has been found to result precisely from the law of gravitation. Not only is the line of intersection of the plane of the moon's orbit with that of the earth constantly changing, but theory, as well as observation, has ascertained that a series of changes are equally progressing in the angles of inclination of these two planes. The limits are narrow, but the oscillations are unceasing, complicating more and more the relative motions of these two remarkable bodies.

In the physical examination of the revolution of the planetary orbs by the application of the law of gravitation, the general features of the investigation are greatly simplified by the fact that the planets and satellites may be regarded as spherical bodies, and may in general be treated as though their entire mass were condensed into a material heavy point situated at their centre. While this statement is true in its broader application to the theory of planetary perturbations, or even in the theory of the sun's action on the planets, especially the more distant ones, it is by no means to be admitted, when we come to a critical examination of the figures of the planets, and the influence exerted by these figures on their near satellites.

In case the earth had been created an exact sphere, and had been projected in its orbit without any rotation on an axis, then would its globular figure have remained without sensible change. But as it revolves swiftly on its axis, the laws of motion and gravitation come in to modify the figure of the earth, and to change it from an exact spherical figure to one which is flattened at the poles and protuberant at the equator. Newton's sagacity detected this result as a necessary consequence of the action of gravitation, and he actually computed the figure of the earth from theory, long before any observation or measurement had created a suspicion that its ℓ, m was other than spherical. The truly wonderful train of consequences flowing from the spheroidal form of the earth gives to this subject a high interest, and demands as close an examination of its principal features as the nature of our investigations will permit.

Give to the earth, then, an exactly spherical form, and a diameter of 8000 miles, with a rotation on an axis once in twenty-four hours, and let us critically examine the consequences. A particle of matter situated on the equator is 4,000 miles from the earth's axis, and since it passes over the circumference of a circle whose radius is 4,000 miles, it will move with a velocity of about one thousand miles an hour. As we recede from the equator towards the poles, either north or south, the particles revolve at the extremities of radii constantly growing shorter and shorter, until finally at the exact pole there is no motion whatever. But in every revolving body, a centrifugal force is generated—a tendency or disposition to fly from the axis of rotation in a plane perpendicular to this axis.

Such is the power of this centrifugal force, that if it were possible to make the earth rotate *seventeen times* in twenty-four, instead of once, bodies at the equator would be lifted up by the centrifugal force, and the attraction of gravitation would be counterpoised, if not absolutely overcome. The force of gravity exerts its power in directions passing nearly through the centre of the earth, while the centrifugal force is always exerted in a direction perpendicular to the axis of rotation. The consequence is manifest, that these two forces cannot counterpoise each other, except in their action on particles situated on the equator of the revolving body. Let us consider the condition of a particle situated anywhere between the equator and the pole, and free to move under the joint action of these two forces.

In order that such a particle may be held in equilibrium, the two forces must act on the same straight line, and in opposite directions. This is not the case in question, for gravity draws the particle to the centre of the earth, while the centrifugal force urges it from the axis in a plane perpendicular to that axis. The direction of these two forces is inclined under an angle which is nothing at the equator, and increases from the equator to the poles. But the effect produced by the centrifugal force may always be obtained by the joint action of two forces, the one directed to the centre of the earth, the other tangent to the earth's surface. Substituting these two forces for the centrifugal force, we perceive that the partial force directed towards the earth's centre is destroyed by gravitation, while the tangential force exerts its full power to move the particle towards the equator of the earth.

This being understood, it is manifest that as particles are coming constantly from both poles towards the equator, that a change of figure in the earth must be effected. It becomes protuberant at the equator, and is flattened at the poles.

The question now arises, whether there be any limit to this change of figure. In case the velocity of rotation continues undiminished, is there not reason to fear that the earth will grow more and more protuberant at the equator, heaping up the matter higher and higher, till the figure of the earth be destroyed, and its surface rendered uninhabitable? Theory has answered this important question; and it has been fully demonstrated that the figure of the earth cannot pass a limit, which it has even now actually attained, and its present form will not change, from the action of the centrifugal force, in millions of years. A condition of equilibrium has been attained, and all further change is at an end. Indeed, if we examine carefully the subject, we may readily perceive, from the nature of the forces, and the conditions of the problem, that such a result might have been anticipated. As the earth grows more protuberant, changing from the spherical form, the particles must be heaved up the side of this elevated ridge which belts the earth around the equatorial regions, and finally the resistance they meet from the elevation they are obliged to overcome, is quite equal to the moving force, and the two destroy each other. This point attained, equilibrium ensues, and further change becomes impossible.

Such is the beautiful order of nature, such the admirable arrangement for stability and perpetuity, everywhere manifested, that the thought constantly comes to the mind that divine wisdom alone could have framed so admirable a system.

But the question may here arise, is this a mere theoretic result? Has observation confirmed the theoretic figure? I answer that observations, the most numerous and diversified, have all united their harmonious testimony to the truth of these beautiful results. In executing exact measures of the degrees of a meridian passing through the poles round the earth, the length of the degree is found to increase from the equator towards the poles, showing that the curvature is more flattened as we recede from the equator. But a more delicate proof is found in the vibrations of the pendulum. A pendulum of a given length will vibrate with a velocity precisely proportioned to the intensity of the force of gravity which operates on it. But the intensity of gravity decreases as the square of the distance from the centre increases, so that it is manifest that the force of gravity is less at the equator than at the poles, in case the surface at the equator is farther from the centre than at the poles, which is the fact asserted by theory.

This being understood, we are prepared to determine the exact figure of the earth, by transporting a pendulum of given length from the equator to different latitudes north and south.—The number of vibrations in one hour being accurately counted at the equator, as we recede north or south, will determine with certainty whether we are approaching to, or going farther from the earth's centre. These experiments have actually been

UNIVERSAL GRAVITATION.

performed, and with the most satisfactory results. The number of vibrations in an hour increases the farther we go north or south, and in a ratio giving the strongest confirmation to the truth of the earth's figure derived from the theoretic investigations—each combining to show that the polar diameter of the earth is but 7,898 miles, while the equatorial diameter is 7,924 miles, producing a sort of ridge or belt around the equatorial regions, rising about thirteen miles above the general spherical surface described about the polar axis as a diameter.

More than two thousand years have passed away since a discovery was made, showing that the sun's path among the fixed stars was slowly changing.—The point at which it crossed the equatorial line, and which for ages had been regarded as fixed, was finally detected to have a slow retrograde motion, producing the *precession of the equinoxes*. The fact was received but, no depth of penetration, no stretch of intellectual vigor could divine the cause of this inexplicable change. Another fact was revealed about the same time. It was found by attentive examination, that the north pole of the heavens, the point in which the prolongation of the earth's axis pierces the celestial sphere, was actually changing, by slow degrees, its place among the fixed stars. The bright star which, in former ages, had marked the place of the pole, and whose circle of diurnal revolution was scarcely to be perceived from its smallness, as centuries slowly glided by, was increasing its distance from the pole, gradually describing round it a circle of greater radius. An attentive examination of the stars near the pole soon demonstrated the fact, that it was an actual motion of the pole, and not of the stars in its neighborhood.

Now, incredible as this statement may appear, modern science has traced these phenomena, the revolution of the equinoctial point, and the movement of the north pole of the heavens, to a common origin, and has demonstrated, in the clearest manner, that they are both consequences of the spheroidal figure of the earth, which we have just examined. It is not my design to enter into an elaborate investigation of this wonderful subject, but, in accordance with the plan already announced, I cannot leave you with a mere announcement of a truth so startling, without some effort to explain how this may be. The subject is difficult, but favored by your close attention, I do not despair of rendering it approximately intelligible.

Let us conceive the earth's axis to be a solid bar of iron driven through the centre of the earth, coming out at the poles, and extending indefinitely towards the sphere of the fixed stars. Now turn this axis up until it stands perpendicular to the plane of the orbit in which the earth revolves round the sun. Then do the equator and ecliptic exactly coincide, and the fixed stars are at a distance nearly infinite, the point in which the earth's axis prolonged pierces the heavens will appear stationary, so far as the revolution of the earth in its orbit is concerned. Now if this iron axle could be grasped by some giant hand, and drawn away from its upright or perpendicular position, the solid earth would turn with it, and the equator, ceasing to coincide with the ecliptic or plane of the earth's orbit, comes to be inclined to it, under an angle precisely equal to the angle through which the axis has been inclined. It is thus seen that no change can be wrought on the position of the axis, that does not involve a corresponding change in the whole earth, and especially in the plane of the equator, which must ever remain perpendicular to the axis in all its positions.

The reverse of this proposition is equally manifest. If the solid earth be seized at the equator, and be turned up or down, the axis will participate in this movement, and its changes will exhibit itself in the changed position of the point in which it meets the celestial sphere. One step more, and the difficulty is surmounted.—Conceive a flat wheel of wood floating on still water. Through its centre pass an axle which stands perpendicular to the surface of the wheel and water. So long as the wheel floats level, the axle stands erect, but in case the north half of the wheel is tilted down under the water, the south half at the same time rising out of the water, the axis will tilt towards the *north*. Bring the wheel again to its level position. Now plunge the eastern portion of the wheel below the surface.—The axis now is tilted towards the *east*.—The experiment is simple, and shows that, in case the successive portions of the wheel be submerged, the axis will always be tilted towards the point which goes under first. To reverse the experiment: in case we take hold of the axle and turn it east it sinks the eastern half of the wheel below the surface of the water, while the western half is raised out of it, and then in case we make the upper extremity of the axis follow round the circumference of a circle whose surface is parallel to that of the water, and whose centre is exactly above the centre of the wheel, it will be seen that, as the axle moves round, successive portions of the wheel are sumerged, until finally the water line will have divided the wheel into all its successive halves, and will have successively coincided with every possible diameter of the wheel.

Now for the application. The level surface of the water is the level plane of the earth's orbit, the wheel is the earth's equator, and the axle is the earth's axis of rotation. One-half of the equator is constantly submerged below the plane of the ecliptic—the other half rises above it. But the water line, or the intersection of these two planes, the equinoctial line, cannot remain fixed in the same line. A power does seize the equator and plunge successive halves of it beneath the plane of the ecliptic, changing perpetually the water line, until finally each half in succession, into which all its diameters can be divided, are sunk below the surface, or plane of the ecliptic, thus causing the earth's axis to tilt over towards the portions successively submerged, until it finally sweeps entirely round and comes to resume its first position.

But do you now demand what power seizes the earth's protuberant equator, and tilts it successively towards every point of the compass? I answer that the power is lodged in the sun and moon, and it is their combined action which works out these wonderful results. In case the sun and moon were so situated as always to be in the plane of the earth's equator, then they would have no power to change the position of the equator. But we know that th y are not in this plane, except when passing through it, and are found som times on th north and sometimes on the south side of it. Wherever either of them may be, the nearest half of the redundant matter about the earth's equator will be more forcibly attracted than the remote half, and the equator will be tilted towards the attracting body, and the axis of the earth will follow the movement of the equator to which it is firmly fixed.

Thus does the earth's whole solid mass sway to the motion of the ring of matter heaped up around the equator, delicately and beautifully sensitive to all the changes in th relative places of the sun and moon. Neither the earth nor its axis are ever, for one moment, released from the action of these remote bodies. However slight the effects, however varied in action, oscillating to every point within certain prescribed limits, the stability is preserved, and the final effect is a small retrograde motion of the equinox at the end of every year, and a slight change in the place of the pole of the heavens.

But there is no isolated matter in the universe.—Every particle of matter attracts every other particle of matter, and it is impossible for the sun and moon to exert any influence on the equatorial ring f matter which belongs to our globe, without feeling, in their turn the reaction of this ring on themselves.—The remote and ponderous sun may, in consequence of of its vast size and distance, escape from any effect capable of being detected by observation. But this is not the case with the moon. Her proximity to the earth and diminutive mass, render her peculiarly sensitive to the influ nce of the redundant matter at the earth's equator, and as her attraction tilts the plane of the earth's equator, so does the equatorial ring tilt the plane of the moon's orbit. These effects have been accurately observed and measured, and strange to relate, their exact values have exhibited the figure which belongs to the earth with far greater precision than can be obtained from measures on its surface. We may even go farther, for such is the intimate relationship between the earth and its attendant satellite, that there is scarcely a question can be asked with reference to the one, that is not answered by the other.

If we demand the weight of the earth when compared with the sun, the moon answers. If the excess of the equatorial diameter of the earth over the polar be required, the moon answers. If the homogenity of the interior of the solid earth be required, the moon replies, If the thickness of the earth's crust is sought for, question the moon, and the answer comes. If you would know the sun's distance from the earth, ask the moon. If the permanency of the axis of rotation be required, ask the moon, and she alone yields a satisfactory reply. Finally, if curiosity leads us to inquire whether the length of the day and night, the revolution of the earth on its axis, be uniform, or whether it may not have changed by a single second in a thousand years, we go to the moon for an answer, and in each and every instance her replies to all the profound and mysterious questions are clear and satisfactory. How wonderful the structure of the universe! How gigantic the power of the human intellect! If all the stars of heaven were struck from existence; if every planet and satellite which the eye and the telescope descry, inside and beyond the earth's orbit, were swept away forever, and the sun, earth and moon alone remained for the study of man, and as evidences of the being and wisdom of God, in the exquisite adjustme ts of this system, in the reciprocal influences of its three bodies, in their vast cycles of cnfiguration, in their relatives masses, magnitudes, distances, motions, and perturbations, there would remain themes suffici nt for the exercise of the most exalted genius, and proof of the being of God, so clear a d positive, that no sane mind could comprehend it and disbelieve.

LECTURE VI.

THE STABILITY OF THE PLANETARY SYSTEM.

WHEN, by the application of a single great law, the mind had succeeded in resolving the difficult problems presented by the motions of the earth and its satellite, the moon, it rose to the examination of the higher and more complicated questions of the stability of the entire system of planets, satellites, and comets, which are found to pursue their courses round the sun. The number of bodies involved in this investigation, their magnitudes and vast periods of revolution, their great distances from the observer, and the exceeding delicacy of the required observations, combined with the high interest which attaches itself to the final result, have united to render this investigation the most wonderful which has ever employed the energies of the human mind.

To comprehend the dignity and importance of this great subject, let us rapidly survey the system, and moving outward to its known boundaries, mark the number and variety of worlds involved in the investigation. Beginning, then, at the great centre, the grand controlling orb, the sun, we find its magnitude such as greatly to exceed the combined masses of all its attendant planets. Indeed, if these could all be arranged in a straight line on the same side of the sun, so that their joint effect might be exerted on that body, the centre of gravity of the entire system thus located, would scarcely fall beyond the limits of the sun's surface. At a mean distance of 36,000,000 of miles from the sun we meet the nearest planet, Mercury, revolving in an orbit of considerable eccentricity, and completing its circuit around the sun in a period of about eighty-eight of our days. This world has a diameter of only 3,140 miles, and is the smallest of the old planets. Pursuing our journey, at a distance of 68,000,000 of miles from the sun, we cross the orbit of the planet Venus. Her magnitude is nearly equal to that of the earth. Her diameter is 7,700 miles, and the length of her year is nearly 225 of our days. The next planet we meet is the earth, whose mean distance from the sun is 95,000,000 of miles. The peculiarities which mark its movements and those of its satellite, have been already discussed. Leaving the earth, and continuing our journey outward, we cross the orbit of Mars, at a mean distance from the sun of 142,000,000 of miles. This planet is 4,100 miles in diameter, and performs its revolution around the sun in about 687 days, in an orbit but little inclined to the plane of the ecliptic. Its features, as we shall see hereafter, are more nearly like those of the earth than any other planet. Beyond the orbit of Mars, and at a mean distance from the sun of about 250,000,000 of miles, we encounter a group of small planets, eight in number, presenting an anomaly in the system, and entirely different from anything elsewhere to be found. These little planets are called asteroids. Their orbits are, in general, more eccentric, and more inclined to the ecliptic, than those of the other planets; but the most remarkable fact is this:—that their orbits are so nearly equal in size, that when projected on a common plane, they are not enclosed, the one within the other, but actually cross each other.

We shall return to an examination of these wonderful objects hereafter. At a mean distance of 485,000,000 of miles from the sun, we cross the orbit of Jupiter, the largest and most magnificent of all the planets. His diameter is nearly 90,000 miles. He is attended by four moons, and performs his revolution round the sun in a period of nearly twelve years.— Leaving this vast world, and continuing our journey to a distance of 890,000,000 of miles from the sun, we cross the orbit of Saturn, the most wonderful of all the planets. His diameter is 76,068 miles, and he sweeps round the sun in a period of nearly twenty-nine and a half years. He is surrounded by several broad concentric rings, and is accompanied by no fewer than seven satellites or moons. The interplanetary spaces we perceive are rapidly increasing. The orbit of Uranus is crossed at a mean distance from the sun of 1,800,000,000 of miles. His diameter is 35,000 miles, and his period of revolution amounts to rather more than eighty-four of our years. He is attended by six moons, and pursues his journeys at a slower rate than any of the interior planets. Leaving this planet, we reach the known boundary of the planetary system, at a distance of about 3,000,000,000 of miles from the sun. Here revolves the last discovered planet, Neptune, attended by one, probably by two moons, and completing his vast circuit about the sun in a period of one hundred and sixty-four of our years.—His diameter is eight times greater than the earth's, and he contains an amount of matter sufficient to form one hundred and twenty-five worlds such as ours.

Here we reach the known limit of the planetary worlds, and standing at this remote point and looking back towards the sun, the keenest vision of man could not descry more than one solitary planet along the line we have traversed. The distance is so great, that even Saturn and Jupiter are utterly invisible, and the sun himself has shrunk to be scarcely greater than a fixed star.

There are certain great characteristics which distinguish this entire scheme of worlds. They are all nearly globular—they all revolve on axes—their orbits are all nearly circular—they all revolve in the same direction around the sun—the planes of their orbits are but slightly inclined to each other, and their moons follow the same general laws. With a knowledge of these general facts, it is proposed to trace the reciprocal influences of all these revolving worlds, and to learn, if it be possible whether this vast scheme has been so constructed as to endure while time shall last, or whether the elements of its final dissolution are not contained within itself, either causing the planets, one by one, to drop into the sun, or to recede from this great centre, released from its influence, to pursue their lawless orbits through unknown regions of space.

Before proceeding to the investigation of the great problem of the stability of the universe, let us examine how far the law of gravitation extends its influence over the bodies which are united in the solar system. A broad and distinct line must be drawn between those phenomena, for which gravitation must render a satisfactory account, and those other phenomena, for which it is in no wise responsible. In the solar system we find, for example, that all the planets revolve in the same direction around the sun, in orbits slightly elliptical, and in planes but little inclined to each other. Neither of these three peculiarities is in any way traceable to the law of gravitation.

Start a planet in its career, and, no matter what be the eccentricity of its orbit, the direction of its movement, or the inclination of the plane in which it pursues its journey, once projected, it falls under the empire of gravitation, and ever after, this law is accountable for all its movements. We are not, therefore, to regard the remarkable constitution of the solar system as a result of any of the known laws of nature.

If the sun were created, and the planetary worlds formed and placed at the disposal of a being possessed of less than infinite wisdom, and he were required so to locate them in space, and to project them in orbits, such that their revolutions should be eternal, even with the assistance of the known laws of motion and gravitation, this finite being would fail to construct his required system.

Let it be remembered, that each and every one of these bodies exerts an influence upon all the others. There is no isolated object in the system. Planet sways planet, and satellite bends the orbit of satellite, until the primitive curves described, lose the simplicity of their character, and perturbations arise, which may end in absolute destruction. There is no chance work in the construction of our mighty system. Every planet has been weighed and poised, and placed precisely where it should be. If it were possible to drag Jupiter from its orbit, and cause him to change places with the planet Venus, this interchange of orbits would be fatal to the stability of the entire system. In contemplating the delicacy and complexity of the adjustment of the planetary worlds, the mind cannot fail to recognize the fact that, in all this intricate balancing, there is a higher object to be gained than the mere perpetuity of the system.

If stability had been the sole object, it might have been gained by a far simpler arrangement. If God had so constituted matter that the sun might have attracted the planets, while these should exert no influence over each other—that the planets might have attracted their satellites, while these were free from their reciprocal influences—then, indeed, a system would have been formed, whose movements would have been eternal, and whose stability would have been independent of the relative positions of the worlds, and the character of their orbits. Give to them but space enough in which to perform their revolutions around the sun, so that no collisions might occur, freed from this only danger, every planet, and every satellite, will pursue the same undeviating track throughout the ceaseless ages of eternity.

If this statement be true, it may be demanded, why such a system was not adopted. It is impossible for us to assign all the reasons which led to the adoption of the present complicated system. Of one thing, however, we are certain:—If God designed that in the heavens his glory and his wisdom should be declared, and that in the study of his mighty works, his intelligent creatures should rise higher and higher towards his eternal throne, then, indeed, has the present system been admirably constituted for the accomplishment of this grand design. To have acquired a knowledge of a system constituted of independent planets, free from all mutual perturbations, would have required scarcely no effort to the mind, when compared with that put forth in the investigation of the present complex construction of the planetary system. The mind would have lost the opportunity of achieving its greatest triumphs, while the evidence of infinite wisdom displayed in the arrangement and counterpoising of the present system would have been lost forever. There is one other thought which here suggests itself with so much force that I cannot turn away from it. We speak of gravitation as some inherent quality or property of matter, as though matter could not exist in case it were deprived of this quality. This is, however, a false idea. Matter might have existed independent of any quality which should cause distant globes to influence each other.—This force called gravitation, even admitting that it must have an existence, no special law of its action could have forced itself on matter to the exclusion of all other laws. Why does this force diminish as the *square* of the distance at which it operates, *increases?* There are almost an infinite number of laws, according to which an attraction might have exerted itself, but there is no one which would have rendered the planets fit abodes for sentient beings, such as now dwell on them, and which would at the same time have guaranteed the perpetuity of the system. Admitting, then, that matter cannot be matter, without exerting some influence on all other matter, (which I am unwilling to admit), in the selection of the law of the inverse square of the distance, there is the strongest evidence of design.

If we rise above the law of gravitation to the Great Author of nature, and regard the laws of motion and of gravitation as nothing more than the uniform expressions of his will, we perceive at once the impossibility of constructing the universe in such manner that the sun should attract the planets, without these attracting each other; or that the planets should attract their satellites without, in turn, being reciprocally influenced by their satellites; for this would be equivalent to saying that the will of the same Almighty Being should exert itself, and not exert itself, at the same moment, which is impossible. As there is but one God, so there is but one kind of matter, governed by one law, applied by infinite wisdom to the formation of suns and systems without number, crowding the illimitable regions of space, all moving harmoniously, fulfilling their high destiny, and all sustained by the single arm of divine Omnipotence.

We now proceed to an examination of the great question, Is the system of worlds by which we are surrounded, and of which our earth and its moon form a part, so constructed that, under the operation of the known laws of nature, it shall forever endure, without ever passing certain narrow limits of change, which do not any way involve its stability?

It is well known that the planets revolve in elliptical orbits of small eccentricity—that under the action of the primitive impulse by which they were projected in their orbits, they would have moved off in a straight line, with a velocity proportioned to the intensity of the impulse, and which would have endured forever; but being seized by the central attraction of the sun, at the moment of starting in their career, the joint action of these two forces, bends the planet from its straight direction, and cause it to commence a curvilinear path, which carries it round the sun.

The question which first presents itself is this :—If the central force lodged in the sun has the power to cause a planet to diverge from the straight line in which, but for this, it would have moved—if it draw it into a curved path, will not this central force, which is ever active, finally overcome, entirely, the impulsive force originally given to the planet, draw it closer and closer to the sun in each successive revolution, in a spiral orbit, until, finally, the planet shall fall into the sun, and be destroyed forever? This question arises independent of the extraneous influence which the planets exert over each other. It refers to a solitary globe revolving around the sun, under the influence of a central force which varies its action as does the law of gravitation. The problem has been submitted to the most rigorous mathematical examination, and a result has been obtained which settles the question in the most absolute manner. The amount by which the central force, in a moment of time, overcomes the effect produced by the primitive impulse, is a quantity, *infinitely small* and of the *second order*. If it were found to be *infinitely small* in each moment of time, then might it accumulate so that, at the end of a vast period, it might become finite and appreciable. But because it is of the *second order* of *infinitely small quantities*, before it can become an infinitely small quantity of the *first* order, a period equal to infinite ages must roll by, and to make a finite appreciable quantity out of this, an infinite cycle of years must roll round an infinite number of times!

Such is the answer given by analysis to this wonderful question. "Is there no change?" demands the astronomer. "Yes," answers the all-seeing analysis. "When will it become appreciable?" asks the astronomer. "At the end of a period infinitely long, repeated an infinite number of times," is the reply.

Having settled this important question, it remains now to examine whether the mutual attractions of the planets on each other may not, in the end, change permanently the form of their orbits, and lead, ultimately, to the destruction of the system. To comprehend more readily the nature of the examination, let us review the points involved in the permanency of our orbit.

Take, for example, our own planet, the earth. It now revolves in an elliptic orbit, whose magnitude is determined by the length of its longer axis, and by its eccentricity. These elements are readily deduced from observation. If it were possible to construct this orbit of some material, like wire, which would permit us to take it up and locate it in space at will, to enable us to give it the position now occupied by the actual orbit of the earth, we must first carry its focus to the sun's centre; we must then turn its longer axis around this centre as a fixed point, until the nearest vertex of the wire orbit shall fall upon that point of the earth's orbit which is at this time nearest to the sun. Having accomplished this, the axes will coincide in their entire length, and to make the orbits coincident, we must revolve the artificial one around the now common axis, until its plane shall fall upon the actual orbit of the earth.

If, now, change should ever come, in the absolute coincidence of these two orbits, regarding the iron one as fixed and permanent, the orbit of nature may vary from it in any one or all of the following ways:— First. The natural orbit, all other things remaining the same, may leave the fixed orbit by a variation of eccentricity;—that is, it may become more or less nearly circular. Second. The planes of the orbits remaining coincident, the curves may separate from each other, in consequence of an angular movement of the longer axis of the natural orbit, by means of which the vertex of the natural curve shall be carried to the right or to the left of the vertex of the fixed one. Third. While these causes are operating to produce change, an increase of deviation may be occasioned by the fact that the two planes may become inclined to each other, thus causing the natural orbit to lie partly above and partly below the fixed one.—These, then, are the several ways in which the orbits of the planets may change; and to settle the question of stability, we must ascertain whether these changes actually exist, and whether any of them, in case they do exist, and are progressing constantly in the same direction, will ever prove fatal to the permanency of the system, finally accomplishing its absolute destruction, or rendering it unfit for the sustentation of that life which now exists upon the planet.

By a close examination of this great subject, both theoretically and practically, it is found that the system is so constituted, that not a single planet or satellite revolves in an orbit absolutely invariable. Theory demonstrates that such changes must exist, and observation confirms this great truth, by showing that they actually do exist.

Draw, in imagination, a straight line from the sun's centre, through the perihelion, or nearest point to the sun of the earth's orbit, and let it be extended to the outermost limits of the entire system. On this line locate the perihelion points of the orbits of all the planets, in these points fix the planets themselves. They are now all on the same side of the sun, the longer axes of their orbits are in the same direction, and they are all located at their nearest distance, from the sun, or in perihelion. The planes of the orbits are inclined to each other under their proper angles, and they all intersect in a common line of nodes passing through the sun's centre. Now give the entire group of planets their primitive impulse, and at the same instant they start in their respective orbits round the sun. Now, in case no perturbations existed, the perihelion points, the inclinations, and the lines of nodes, would remain fixed forever, and although millions of years might pass away before the planets would again resume their primitive position with reference to each other, yet the time would come when a final restoration would be effected.

At the end of 164 years, Neptune will have completed its revolution round the sun, and will return to its starting point. All the other planets will have performed several revolutions, but each, on reaching the point of departure, will find the perihelion of its orbit changed in position, the inclination altered and the line of nodes shifted. These changes continue until the longer axes of the orbits, which once coincided, radiate from the sun in all directions. The lines of nodes, once common, now diverge under all angles, the inclinations increasing or decreasing, and even the figures of the orbits undergoing constant mutation; and the grand question arises, whether these changes, no matter how slow, are ever to continue progressing in the same direction, until all the original features of the system shall be effaced, and the possibility of return to the primitive condition destroyed forever.

Such a problem would seem to be far too deep and complicated ever to be grasped by the human intellect. It is true that no single mind was able to accomplish its complete solution, but the advance made by one has been steadily increased by another, until, finally, not a question remains unanswered. The solution is complete, yielding results of the most wonderful character.

We shall examine this great problem in detail, and commence with the figure of the orbit of any planet, our earth, for example.

STRUCTURE OF THE UNIVERSE.

The amount of heat received from the sun by the earth depends, other things being the same, on the minor axis of its ecliptic orbit. Any change in the eccentricity operates directly to increase or decrease the shorter axis, and consequently to increase or decrease the mean annual amount of heat received from the sun. Now we know that animal and vegetable life is adjusted in such way that it requires almost exact uniformity in the mean annual amount of heat which it shall enjoy. An increase or decrease of two or three degrees in temperature would make an entire revolution in the animals and plants belonging to the region experiencing such a change. If, then, it be true that the eccentricity of the earth's orbit is actually changing, under the combined action of the other planets, may this change continue so far as to subvert the order of nature on its surface?— This question has been answered in the most satisfactory manner.

It is found that the greater axes of the planetary orbits are subjected to slight and temporary variations, returning, in comparatively short periods, to their primitive values. This important fact guarantees the permanency of the periodic times, so that it becomes possible to deduce, with the utmost precision, the periodic times of the planets, from the mean of a large number of revolutions. That of the earth is now so accurately known, and so absolutely invariable, that we know what it will be a million of years hence, should the system remain as it now is, as perfectly as at the present moment. But neither of these elements secures the stability of the eccentricity, or of the minor axis. Lagrange, however, demonstrated a relation between the masses of the planets, their major axes and eccentricities, such, that while the masses remain constant, and the axes invariable, their eccentricity can only vary its value through extremely narrow limits. These limits have been assigned, beyond which the change can never pass, and within these narrow bounds we find the orbits of all the planets slowly vibrating backward and forward, in periods which actually stun the imagination.

This remarkable law for the preservation of the system would not hold in any other organization. It demands orbits nearly circular, with planes nearly coincident, with periodic times related as are those of the planets, and the planets themselves located as they actually are. No interchange of orbits is admissible; but, constituted as the system now is, the perpetuity is absolutely certain, so far as the change of eccentricity is concerned.

Let us now examine the changes which affect the position of the major axis in its own plane. The perihelion of every orbit is found to be slowly advancing. Nor is this advance ever to be changed into a retrograde motion. The movement is ever progressive in the same direction, and the perihelion points of all the orbits are slowly sweeping round the sun.— That of the earth's orbit accomplishes its revolution in *one hundred and eleven thousand years!* How wonderful the fact, that such discoveries should be made by man, whose entire life is but a minute fraction of these vast periods of time!

Owing to a retrograde motion in the vernal equinox, carrying it around in the opposite direction in 25,868 years, the perihelion and equinox pass each other once in 20,984 years. Knowing their relative positions at this moment, and their rates of motion it is easy to compute the time of their coincidence. Their last coincidence took place 4,089 years before the Christian era, or about the epoch usually assigned for the creation of man. The effect of the coincidence of the perihelion with the vernal equinox, is to cause an exact equality in the length of spring and summer, compared with autumn and winter. In other language, the sun will occupy exactly half a year in passing from the vernal to the autumnal equinox, and the other half in moving from the autumnal to the vernal equinox.

At present, the line of equinoxes divides the earth's elliptic orbit into two unequal portions. The smaller part is passed over in the fall and winter, causing the earth to be nearer the sun at this season than in summer, and making a difference in the length of the two principal seasons, summer and winter, of some seventeen and a half days. This inequality, which is now in favor of summer, will eventually be destroyed, and the time will come when the earth will be farthest from the sun during the winter, and nearest in the summer. But at the end of a great cycle of more than 20,000 years, all the changes will have been gone through, and, in this respect, a complete compensation and restoration will have been effected.

This epoch of subordinate restoration will find the perihelion of the earth's orbit located in space far distant from the point primitively occupied. Five of these grand revolutions of 20,984 years must roll round before the slow movement of the perihelion shall bring it back to its starting point. 110,000 years will then restore the axis of the earth's orbit, and the equinoctial line, nearly to their relative positions to each other, and to the same region of absolute space occupied at the beginning of this grand cycle.

If, now, we direct our attention to the other planets, we find their perihelion points all slowly advancing in the same direction. That of the orbit of Jupiter performs its revolution round the sun in 186,207 years, while the perihelion of Mercury's orbit occupies more than 200,000 years in completing its circuit round the sun. To effect a complete restoration of the planetary orbits to their original position, with reference to their perihelion points, will require a grand compound cycle, amounting to millions of years. Yet the time will come when all the orbits will come again to their primitive positions, to start once more on their ceaseless journeys.

In the changes of the eccentricities, it will be remembered, the stability of the system was involved. Should these changes be ever progressive, no matter how slowly, a time would finally come when the original figure of the orbit would be destroyed, the planet either falling into the sun, or sweeping away into unknown regions of space. But a limit is assigned, beyond which the change can never pass.—Some of the planetary orbits are becoming more circular, others growing more elliptical; but all have their limits fixed. The earth's orbit, for example, should the present rate of decrease of eccentricity continue, in about half a million of years will become an exact circle. There the progressive motion of the changes stops, and it slowly commences to recover its ellipticity. This is not the case with the motions of the perihelia. Their positions are in no way involved in the well being of a planet, or in its capacity to sustain the life which exists on its surface; and since the stability of the system is not endangered by progressive change, it ever continues in the same direction, until the final restoration is effected, by an entire revolution about the sun.

Let us now examine the inclinations of the planetary orbits. Here it is found that there is no guarantee for the stability of the system, provided the angles under which the orbits of the planets are inclined to each other do not remain nearly the same forever.—If changes are found to exist, by which the inclinations are made to increase, without stopping and returning to their primitive condition, then is the perpetuity of the system rendered impossible. Its fair proportions must slowly wear away, the harmony which now prevails be destroyed, and chaos must come again.

Commencing again with the earth, we find that, from the earliest ages, the inclination of the earth's equator to the ecliptic has been decreasing. Since the measure of Eratosthenes, 2,078 years ago, the decrease has amounted to about 23′ 44″, or about half a second every year. Should the decrease continue, in about 85,000 years the equator and ecliptic would coincide, and the order of nature would be entirely changed;—perpetual spring would reign throughout the year, and the seasons would be lost forever. Of this, however, there is no danger. The diminution will reach its limit in a comparatively short time, when the decrease of inclination will change into an increase, and thus slowly rocking backwards and forwards in thousand of years, the seasons shall ever preserve their appointed places, and seed time and harvest shall never fail. These changes of inclination are principally due to the perturbations of Venus, and arising from configurations, will be ultimately entirely compensated.

The angles under which the planetary orbits are inclined to each other are in a constant state of mutation. The orbit of Jupiter at this time forms an angle with the ecliptic of 4,731 seconds, and this angle is decreasing at such a rate that, in about 20,000 years the planes would actually coincide. This would not effect the well being of the planets or the stability of the system, but should the same change now continue, the angle between the orbits might finally come to fix them even at right angles to each other, and a subversion of the present system would result.

A profound investigation of the problem of the planetary inclinations, accomplished by Lagrange, resulted in the demonstration of a relation between the masses of the planets, the principal axes of their orbits, and the inclinations, such that, although the angles of inclination may vary, the limits are narrow, and they are all found slowly to oscillate about their mean positions, never passing the prescribed limits, and securing, in this particular, the perpetuity of the system.

Here, again, we are presented with the remarkable fact, that whenever mutation involves stability, this mutation is of a compensatory character, always returning upon itself, and, in the long run, correcting its own effects. If all this mighty system was organized by chance, how happens it that the angular motions of the perihelia of the planetary orbits are ever progressive, while the angular motions of the planes of the orbits are vibrating? Design, positive and conspicuous, is written all over the system, in characters from which there is no escape.

We now proceed to an examination of the lines in which the planes of the planetary orbits cut each other, or the lines in which they intersect a fixed plane. These are called the *lines* of *nodes*. They all pass through the sun's centre, and, in case they ever were coincident, they now radiate from a common point in all directions.

Here is an element in no degree involving in its value the stability of the system, and from analogy we already begin to anticipate that its changes, whatever they may be, will probably progress always in the same direction. This is actually the case. The nodes of the planetary orbits are all slowly retrograding on a fixed plane, and in vast periods, amounting to thousands of years, accomplish revolutions, which, in the end, return them to their primitive positions.

Thus are we led to the following results. Of the two elements which fix the magnitude of the planetary orbits, the principal axes, and the eccentricity, the axes remain invariable, while the eccentricity oscillates between narrow and fixed limits. In the long run, therefore, the magnitudes of the orbits are preserved.

Of the three elements which give *position* to the planetary orbits, viz: the place of the perihelion, the lines of nodes, and the inclinations, the two first ever vary in the same direction, and accomplish their restoration at the end of vast periods of revolution, while the inclinations vibrate between narrow and prescribed limits.

One more point, and we close this wonderful investigation. The last question which presents itself is this:—May not the periodic times of the planets be so adjusted to each, as that the results of certain configurations may be ever repeated without any compensation, and thus, by perpetual accumulation, finally effect a destruction of the system?

If the periodic times of two neighboring planets were exact multiples of the same quantity, or if the one was double the other, or in any exact ratio, then the contingency would arise, above alluded to, and there would be perturbations which would remain uncompensated. A near approach to this condition of things actually exists in the system, and gave great trouble to geometers. It was found, on comparing observations, that the mean periods of Jupiter and Saturn were not constant—that one was on the decrease, while the other was on the increase. This discovery seemed to disprove the great demonstration which had fixed as invariable the major axes of the planetary orbits, and guaranteed the stability of the mean motions. It was not until after Laplace had instituted a long and laborious research, that the phenomenon was traced to its true origin, and was found to arise from the near commensurability of the periodic times of Jupiter and Saturn—five of Jupiter's periods being nearly equal to two of Saturn's. In case the equality were exact, it is plain that if the two planets set out from the same straight line drawn from the sun, at the end of a cycle of five of Jupiter's periods, or two of Saturn's, they would be again found in the same relative positions, and whatever effect the one planet had exerted over the other would again be repeated under the same precise circumstances. Hence would arise derangements which would progress in the same direction, and eventually lead to permanent derangement of the system.

But it happens that five of Jupiter's periods are not exactly equal to two of Saturn's, and in this want of equality safety is found. The difference is such, that the point of conjunction of the planets does not fall at the same points of their orbits, but at the end of each cycle is in advance by a few degrees. Thus the conjunction slowly works round the orbits of the planets, and, in the end, the effect produced on one side of the orbit is compensated for on the other, and a mean period of revolution comes out for both planets, which is invariable. In the case of Jupiter and Saturn, the entire compensation is not effected until after a period of nearly a thousand years.

A similar inequality is found to exist between the earth and Venus with a period much shorter, and producing results much less easily observed. In no instance do we find the periods of any two planets in an exact ratio. They are all incommensurable with each other, and in this peculiar arrangement we find the stability of the entire system is secured.

So far, then, as the organization of the great planetary system is concerned, we do not find within itself the elements of its own destruction. Mutation and change are everywhere found—all is in motion—orbits expanding or contracting—their planes rocking up and down—their perihelia and nodes sweeping in opposite directions round the sun,—but the limits of all these changes are fixed;—these limits can never be passed, and at the end of a vast period, amounting to many millions of years, the entire range of fluctuation will have been accomplished, the entire system, planets, orbits, inclinations, eccentricities, perihelia, and nodes, will have regained their original values and places, and the great bell of eternity will have then sounded *one*.

Having reached the grand conclusion of the stability of the system of planets, in their reciprocal influences, and that no element of destruction is found in the organization, we propose next to inquire whether the same features are stamped on the subordinate groups composing the planetary system.—As our limits will not permit us to enter into a full examination of all the subordinate groups, we shall confine our remarks to our own earth and its satellite, Jupiter and his satellites, and to Saturn, his rings and moons, We shall, in this examination, find it praticable to answer, to some extent, the inquiry as to whether either of these systems has received any shock from external causes. We know nothing as to the future. and can, in this particular, only form our conjectures as to what is to be, from what has been.

We commence our inquiry by an examination of two questions, viz :— Is the velocity of rotation of the earth on its axis absolutely invariable? Has the relation between the earth and moon ever been disturbed by any external cause? There is nothing so important to the well-being of our planet and its inhabitants as absolute invariability in the period of its axical rotation. The sidereal day is the great unit of measure for time, and is of the highest consequence in all astronomical investigations. If causes are operating, either to increase or decrease the velocity of rotation, a time will come when the earth will cease to rotate, or else acquire so great a velocity as to destroy its figure, and in the end, scatter its particles in space.

It is difficult to ascertain from theory a perfectly satisfactory answer to the question of the invariable velocity of rotation of the earth, but Laplace has demonstrated that the length of the day has not varied by the hundredth of one second during the last two thousand years—that is, the length of the day is neither greater nor less than it was two thousand years ago by the hundreth of a second. The reasoning leading to this remarkable result is simple, and may be readily comprehended by all. Two thousand years ago, the duration of the moon's period of revolution around the earth was accurately determined, and was expressed in days and parts of a day. The measure of the same period has been accomplished in our own time, and is expressed in days and parts of a day. Now all the causes operating to change the moon's period of revolution are known, and may be applied. When this is done it is found that the moon's period now and two thousand years ago, agree precisely, being accomplished in the same number of days and parts of a day—which would be impossible, if the unit of measure, the day, had varied ever so slightly.

The extraordinary relation existing between the moon's period in her orbit and the time occupied in her axical rotation, gives us the opportunity of ascertaining whether our system has received any external shock. These two periods are so accurately adjusted, that in all respects an exact euqality exists. The moon ever turns the same hemisphere to the earth, and ever will, unless some external cause should arise to disturb the perfect harmony which now reigns. It it not my purpose to explain why it is that this phenomenon exists. I merely desire to state, that this delicate balancing of periods furnishes an admirable evidence that, for several thousands of years, at least, no shock has been received by the earth and its satellite. Steadily have they moved in their orbits, subject only to the influence of causes originating in the constitution of the mighty system of which they constitute a part.

Moving out to a more complex system, we find in the remarkable arrangement of the satellites of Jupiter, a delicate test for the action of sudden and extraneous causes. Here we find the periodic times of the satellites so related, that a thousand periods of the first, added to two thousand periods of the third, will be precisely equal to three thousand periods of the second. This delicate balancing of periods would be destroyed by the action of any external shock, such as might be experienced from the collision of a comet sweeping through the system. Thus far, we know that no disturbance has entered, and a knowledge of facts will now pass down to posterity, which will give the means of ascertaining exactly the influence of all disturbing causes which do not form a part of the great system.

The last subordinate group, and the most extraordinary one to which I will at this time direct your attention, is that of Saturn and his rings. Here we find a delicacy of adjustment and equilibrium far exceeding any thing yet exhibited in our examinations. This great planet is surrounded certainly by two, probably by three immense rings, which are formed of solid matter, in all respects like that constituting the central body. These wonderful appendages are nowhere else to be found, throughout the entire solar system, at least with certainty. Their existence has elsewhere been *suspected*, but around Saturn they are seen with a perfection and distinctness which defies all skepticism as to their actual existence. The diameter of the outer ring is no less than 176,000 miles. Its breadth is 21,000 miles, while its thickness does not exceed one hundred miles. The inner ring is separated from the outer one by a space of about 1,800 miles,—its breadth 34,000 miles, its inner edge being about 20,000 miles from the surface of the planet.—Its thickness is the same as that of the outer ring. These extraordinary objects are rotating in the same direction as the planet, and with a velocity so great that objects on the exterior edge of the outer ring are carried through space with the amazing velocity of nearly 50,000 miles an hour, or nearly fifty times more swiftly than the objects on the earth's equator.

What power of adjustment can secure the stability of these stupendous rings? No solid bond fastens them to the planet—isolated in space, they hold their places, and revolving with incredible velocity around an imaginary axis, they accompany their planet in its mighty orbit round the sun. Such is the exceeding delicacy with which this system is adjusted, that, the slightest external cause once deranging the equilibrium, no readjustment would be effected. The rings would be thrown on the body of the planet, and the system would be destroyed.

To understand the extraordinary character of this system, we will explain a little more fully the three different kinds of equilibrium. The first is called an equilibrium of *instability*, and is exemplified in the effort to balance a rod on the tip of the finger. The slightest deviation from the exact vertical, increases itself constantly, until the equilibrium is destroyed.—In case the same rod be balanced on its centre on the finger, it presents an example of an equilibrium of *indifference;* that is, if it be swayed slightly to the one side or the other, there is no tendency to restore itself, or to increase its deviation.—It remains indifferent to any change. Take the same rod, and suspend it like a pendulum. Now cause it to deviate from the vertical to the right or left, and it returns of itself to the condition of equilibrium. This is an equilibrium of *stability*. We have already seen that this is the kind of equilibrium which exists in the planetary system. There are constant deviations, but a perpetual effort is making to restore the object to its primitive condition.

Now in case the rings of Saturn are homogeneous, equally thick, and exactly concentric with the planet, their equilibrium is one of instability. The smallest derangement would find no restorative power, and would even perpetuate and increase itself, until the system is destroyed. For a long time it was believed that the rings were equally thick, and concentric with the planet, but when it was discovered that such features would produce an equilibrium of instability, and that there existed no guarantee for the permanency of this exquisite system, an analytic examination was made, which led to this singular result, viz :—To change the equilibrium of instability into one of stability, all that is necessary is to make the ring thicker or denser in some parts than in others, and to cause its centre of position to be without the centre of the planet, and to perform around that centre a revolution in a minute orbit. Finding these conditions analytically, it now became a matter of deep interest to ascertain whether these conditions actually existed in nature. The occasional disappearance of the ring, in consequence of its edge being presented to the eye of the observer, gave a capital opportunity of determining whether it was of uniform thickness. On these rare occasions, in the most powerful telescopes, the ring remains visible edgewise, and looks like a slender fibre of silver light drawn across the diameter of the planet. In the gradual wasting away of the two extremities of the ring, it has been remarked, that the one remains visible longer than the other. As the ring is swiftly revolving neither extremity can, in any sense, be regarded as fixed, and hence sometimes the one, some times the other, fades first from the sight. An exactly uniform thickness in the ring would render such a phenomenon impossible, and hence we conclude, that the first condition of stability is fulfilled,—the rings are *not* equally thick throughout.

The micrometer was now applied to detect an eccentricity in the central point of the ring. Recent examinations by Struve and Bessel have settled this question in the most satisfactory manner. The centre of the ring does not coincide with that of the planet, and it is actually performing a revolution around the centre of the planet in a minute orbit, thus forming the second delicate condition of equilibrium. The analogy of the great system is unbroken in the subordinate one. For more than two hundred years have these wonderful circles of light whirled in their rapid career under the eye of man, and freed from all external action, they are so poised that millions of years shall in nowise effect their beautiful organization. Their graceful figures and beautiful light shall greet the eyes of the student of the heavens when ten thousand years shall have rolled away.

Thus do we find that God has built the heavens in wisdom, to declare his glory, and to show forth his handiwork. There are no iron tracks, with bars and bolts, to hold the planets in their orbits. Freely in space they move, ever changing, but never changed ; poised and balancing ; swaying and swayed ; disturbing and disturbed, onward they fly, fulfilling with unerring certainty their mighty cycles. The entire system forms one grand complicated piece of celestial machinery ;—circle within circle ; wheel within wheel ; cycle within cycle :—revolution so swift as to be completed in a few hours; movements so slow that their mighty periods are only counted by millions of years. Are we to believe that the Divine Architect constructed this admirably adjusted system to wear out, and to fall in ruins even before one single revolution of its complex scheme of wheels had been performed? No.—I see the mighty orbits of the planets slowly rocking to and fro, their figures expanding and contracting, their axes revolving in their vast periods; but stability is there. Every change shall wear away, and after sweeping through the grand cycle of cycles, the shall return to its primitive condition of perfection and beauty.

LECTURE VII.

THE DISCOVERY OF NEW PLANETS.

In the earliest ages of the world, the keen vision of the old astronomers had detected the principal members of the planetary system. Even Mercury, which habitually hovers near the sun, and whose light is almost constantly lost in the superior brilliancy of that luminary, did not escape the eagle glance of the primitive students of the stars. For many thousand years no suspicion arose in the mind, as to the existence of other planets, belonging to the great scheme, and which had remained invisible from their immense distance or their, minute dimensions.—Indeed the grand investigations which have recently engaged our attention, the mutation of the planetary orbits, their perpetual oscillations and final restoration, the equilibrium of the whole system, had been prosecuted and completed before the mind gave itself seriously to the contemplation of invisible worlds.

The singularly inquisitive genius of Kepler, over whom analogy seems to have ever played the tyrant, in an examination of the interplanetary spaces, finding these to increase with regularity in proceeding outward from the sun, until reaching the space between Mars and Jupiter, which was out of all proportion too great, conceived the idea that an invisible planet revolved in this space, and thus completed the harmony of the system. The space from the orbit of Mercury to that of Venus is 31,000,000 of miles; from the orbit of Venus to that of the earth is 27,000,000 of miles; from the earth's orbit to that of Mars is 50,000,000 of miles, but between the orbit of Mars and that of Jupiter, there exists the enormous interval of 359,000,000 of miles. The order is again resumed between the orbits of Jupiter and Saturn, and from these slender data Kepler boldly predicted that a time would come when a planet would be found intermediate between the orbits of Mars and Jupiter, whose discovery would establish a regular progression in the interplanetary spaces. For nearly two hundred years this daring speculation was regarded as one of the wild dreams of a great, but visionary mind.

Towards the close of the eighteenth century, when the planetary orbs had been studied with great care, and a comparatively accurate knowledge of their perturbations had been reached, certain unexplained irregularities gave rise to the suspicion that the movements of Saturn might be disturbed by the action of an unknown planet revolving in a vast orbit, remote from, and far beyond that of Saturn. These speculations led to no serious results, and it was only by a fortunate accident that, on the 13th of March, 1781, Sir William Herschel noticed a small star of remarkable appearance, which happened to fall in the field of his telescope. On applying a greater magnifying power, the strange star showed unequivocal symptoms of increased dimensions. Its position among the neighboring stars was noticed with care, and by an examination on the following evening, the stranger was found to have sensibly changed position. A few nights sufficed to establish the fact that the newly discovered body was actually a wandering star, and not for a moment dreaming of the discovery of a new planet, Herschel announced to the world that he had found a remarkable *comet*. Efforts were made to obtain the orbit of the stranger, on the hypothesis, that like those of all the then known comets, it was extremely elongated. Maskelyn and Lexell soon reached the conclusion that no eccentric orbit could possibly represent the motions of the newly discovered star, and on a close and diligent examination, it was at last discovered to be a primary planet, revolving in an orbit nearly circular, and almost coincident with the plane of the ecliptic. Its motion was progressive, like the other planets, and its vast orbit was only completed at the end of eighty-four of our years. Its distance from the sun was found to be no less than 1,800,000,000 of miles, and its dimensions such that out of it might be formed more than eighty worlds as large as the earth.

This great discovery excited the highest interest in the astronomical world. From the earliest ages, the mighty orbit of Saturn had been regarded as forming the boundary of the vast scheme of planets dependent on the sun. Its slow and majestic motion, its great period and distance, and the wonderful magnificence of its rings and moons, seemed to render it a fitting object to guard the frontiers of the mighty system with which it was associated. But the supremacy of Saturn was now gone forever, and its sentinel position was usurped by Uranus, whose grand orbit expanded to twice their original dimensions the boundaries of the solar system. Far sweeping in the depths of space, this new world pursued its solemn journey, flinging back the light of its parent orb, steadily obedient to the great law of universal gravitation, which held the old planets true to their changing orbits.

Another unit in the number of interplanetary spaces was thus given, and the law which might possibly regulate the distances of the planets from the sun was sought after with an interest and perseverance which could not long fail of its reward. No exact progression was indeed discovered, but the following remarkable empirical law was detected by Prof. Bode:

Write the series	0	3	6	12	24	48	96	192, &c.
Add to each term	4	4	4	4	4	4	4	4
The sums are	4	7	10	16	28	52	100	196.

Now if 10 be assumed as the earth's distance from the sun, the other

THE DISCOVERY OF NEW PLANETS.

terms of the series will represent very nearly the distances of the planets thus:

4   7   10   16   28   52   100   196

Mercury, Venus, Earth, Mars, —, Jupiter, Saturn, Uranus. The fifth term in the series is blank, and falls exactly in the enormous interval which exists between the orbits of Mars and Jupiter, precisely where Kepler had predicted a new planet would be found. As early as 1784, three years after the discovery of Uranus, Baron de Zach, struck with the remarkable law of Bode, even went so far as to compute the probable distance and period of the now generally suspected planet. The impression that a new world would be soon be added to the system grew deeper and stronger in the minds of astronomers, until finally, in 1800, at a meeting held at Lilienthal, by six distinguished observers, the subject was discussed with deep earnestness, and it was finally resolved that the long suspected, but yet, undiscovered world, should be made the object of strict and persevering research. The range of the Zodiac was divided into twenty-four parts, and distributed among an equal number of observers, whose duty it was to scrutinize their particular regions, and detect, if possible, any moving body which might show itself among the fixed stars.

In case it were possible to note down, with perfect precision, the relative places and magnitude of all the stars in a given region, any subsequent changes which might occur would be easily recognized. In other language, could a daguerreotype picture of any region in the heavens be made to-night, and at the end of a year another picture of the same region could be taken, by comparing the number of stars in the one picture with that in the second, in case any one had wandered away from its place, or a stranger had come to occupy a place within the limits of the pictured region, it would be an easy matter to ascertain either the lost star, or the newly arrived stranger. Now, although a daguerreotype picture cannot be had, yet by observation, the exact relative positions of all the visible stars may be mapped out, and a picture formed, which shall become the ready means of detecting future changes.

Such was the method of examination adopted by the congress of astronomers assembled at Lilienthal, in 1800. The organization was made.—Baron de Zach was elected president, and Schroeter was chosen perpetual secretary. To those who have paid but little attention to the circumstances under which this strange enterprise was undertaken, nothing can appear more wild and chimerical. To commence a prolonged research for an invisible world, one that no keenness of vision could detect, and which never could be revealed but by telescopic aid, a world whose magnitude was so small that it would not appear so large as a star of the smallest size visible to the naked eye, and one which must be sought out and detected not by its planetary disk, but by its wanderings among thousands of stars, which it in all respects resembled, and from which it could in no wise be distinguished, but by its motion, seemed like a wasting of time and utter throwing away of labor and energy.

Piazzi, of Palermo, in Sicily, was one of the planet searching association, He had already distinguished himself as an eminent and accurate observer and had with indefatigable zeal constructed a most extensive catalogue of the relative places of the fixed stars, and thus, in some sense, anticipated a part of the labor that the research for the suspected planet contemplated. Assisted by his own and by preceding catalogues, he entered on the great work with the energy and zeal which distinguished all his great astronomical efforts. On the evening of the first day of the year 1801, this astronomer had his attention attracted by a small star in the constellation of the Bull, which he took to be one recorded in the catalogue of Mayer; but on examination, it was found not to occupy any place either on Mayer's or his own catalogue. Yet it was so small that it was an easy matter to account for this fact, by its having been overlooked in preceding explorations of the region in which it was found. With intense anxiety the astronomer awaiting the evening of the following night, to settle the great question whether the newly detected star was a fixed or moving body. On the evening of the 2d of January he repaired to his observatory, and so soon as the fading twilight permitted, directed the telescope to the exact point in which, on the preceding evening, his suspicious star had been located. The spot was blank! But another, which was distant 4' in right ascension, and 3½ in declination, which, on the previous night had certainly been vacant, was now gleaming with the bright little object which, on the preceding evening, had so earnestly fixed his attention, and for which he was again so anxiously seeking. Night after night he watched its retrograde motion,—a motion precisely such as it ought to have, in case it were the long desired planet,—until, on the 12th, it became stationary, and then slowly commenced progressing in the order of the signs. Piazzi was unfortunately taken ill; his observations were suspended, and such was the difficulty of intercommunication, that, although he sent intelligence of his discovery to Bode and Orani, associates in the great enterprise, the newly discovered body was already lost in the rays of the sun, before it became possible to renew the train of observations by which its orbit might be made known. Piazzi feared to announce the newly discovered body to be the suspected planet. His observations were few, and he was the only person in the world who had seen it. Bode no sooner received the intelligence of its discovery, than he at once pronounced it to be the long sought planet, and from the scanty materials furnished by Piazzi, Olbers, Burkhart, and Gauss, all computed the elements of its orbits, settled the great fact that it was a superior planet, and that its orbit was included between those of Mars and Jupiter. Some doubt, however, yet rested on the subject, and the disengagement of the planet from the beams of the sun was awaited with the deepest interest.

Several months passed away. Every eye and every telescope was directed to the region in the heavens where the new planet was expected to be found. The most scrutinizing search was made for its rediscovery, but without any success. But for the high reputation of Piazzi, his wellknown accuracy and honesty, doubts would have arisen as to whether he had not been self-deceived, or was intentionally deceiving others. The subject became of deeper and deeper interest. The world began to sneer at a science which could find a body in the heavens, and when forever lose it. We must remember that Piazzi had followed it through only about 4° out of 360° of its orbit, and on this narrow basis a research was to be instituted, having for its object the determination of the exact position which the lost planet must occupy. Gauss, then comparatively a young man, and little known as a computer, had conceived a new method of determining the orbits of comets, from a very few and very closely consecutive observations. Here was an admirable opportunity of giving a practical proof of the power of his new method. The long and intricate calculation was finished, the place of the lost planet determined, the telescope was directed to the spot, and lo! the beautiful little orb flashed once more on the eager gaze of the youthful astronomer. For one entire year had the planet been sought in vain, and but for the powerful analysis of Gauss, nothing but years of persevering toil could have wiped away the reproach which rested on astronomy.

A sufficent number of observations were soon made to reveal the orbitual elements of the planet, now named Ceres. It was found, in all respects, to harmonize in its movements with the older planets, and its orbit filled precisely the blank in the strange empirical law discovered by Bode. The period and distance hypothetically computed from that law sixteen years before, by Baron de Zach, were verified in the most remarkable manner by the actual period and distance of Ceres. Order and beauty now reigned in the planetary system, and a most signal victory had crowned the efforts of astronomical science.

The only remarkable difference betwen the new planet and the old ones, consisted in its minute size, the great obliquity of its orbit, and the dense atmosphere by which it appears to be surrounded. Its diameter is so small as to render its measure next to impossible, and the best practical astronomers differ widely in their results. Sir William Herschel makes its diameter only 163 miles, while Schroeter cannot make it less than ten times that quantity. The mean of these two extremes is probably near the truth.—No satellites have been found in attendance on this minute planet, although Sir William Herschel suspected the existence of two at one time, a suspicion which subsequent observations have not confirmed.

The beautiful order established in the solar system by the discovery of Ceres was a subject of the highest gratification to the whole astronomical world, and especially to those who had been instrumental in reaching this remarkable result. An opportunity had scarcely presented itself for the expression of delight occasioned by this announcement, before all interested were startled by a declaration from Dr. Olbers, of Bremen, that he had found another planet on the evening of the 28th of March, 1802, with a mean distance and periodic time almost identical with those of Ceres. This discovery broke through all the analogies of the solar system, and presented the wonderful anomaly of two planets revolving in such close proximity, that their orbits, projected on the plane of the ecliptic, actually intersected each other.

The new planet was called *Pallas*, and is of a magnitude about equal to that of Ceres. Its orbit is greatly inclined to the plane of the ecliptic, and its eccentricity is very considerable. The existence of these small planets, in such near proximity, for a long while perplexed astronomers. At length Olbers suggested that these minute bodies might be the fragments of a great world, rent asunder by some internal convulsion of sufficient power to produce the terrific result, but of a nature entirely beyond the boundary of conjecture.

Extraordinary as this hypothesis may appear, the results to which it led are not less remarkable. If a world of large size had been actually burst into fragments, it is easy to perceive that these fragments, all darting away in the orbits due to their impulsive forces would start from the same point, and hence would return at different intervals indeed, but would all again pass through the point of space occupied by the parent orb when the convulsion occurred. Having found two of these fragmentary worlds, the point of intersection of their orbits would indicate the region through which the other fragments might be expected to pass, and in which they might possible be discovered. So reasonable did the views of Olbers appear, that his suggestions were immediately acted upon by himself and several distinguished observers, and on the 2d of September, 1804, Mr. Harding, of Lilienthal, while scrutinizing the very region indicated by Olbers, detected a star of the eighth magnitude, which seemed to be a stranger, and was soon recognized to be another small planet, fully agreeing, in all its essential characteristics, with the theory of Olbers. The new world was named Juno, and is remarkable for the eccentricity of its orbit. Its diameter has not been well determined, owing to its minute size. This discovery gave to the theory of Olbers the air of reality, and finding the nodes of the three fragments to lie in the opposite constellations Cetus and Virgo, he prosecuted his researches in these regions with redoubled energy and zeal.

His efforts were not long without their reward. On the 29th of March, 1807, he detected the fourth of his fragments in the constellation Virgo, and very near the point through which he had, for four years, been waiting to see it pass. This was a most wonderful discovery, and almost fixed the stamp of truth upon the most extraordinary theory which had ever been promulgated.

This new asteroid was named Vesta, and for nearly forty years the examinations which were conducted revealed no new fragment, and it began to be regarded as positively ascertained, that all the small bodies revolving in this region had been revealed to the eye.

But on the 8th day of Dec., 1845, Mr. Hencke, of Dreisen announced to the world the discovery of another asteriod, which was named Astrea. Before two years had rolled round, the same indefatigable observer discovered a sixth member in this wonderful group, which was called Hebe. His success induced other observers to undertake a similar examination, and in a very short time the researches of Mr. Hinds of London, were rewarded by the discovery of a seventh and eighth asteroid, which were named *Iris* and *Flora*.

Thus have we no less than eight of these minute worlds, revolving in orbits so nearly aqual, that for weeks and months these miniature orbs may sweep along in space, almost within hail of each other. Let us now return to an examination of the hypothesis of Olbers, that these are the fragments of a world of large size, which once occupied an orbit intermediate between those of Mars and Jupiter.

If any internal convulsion could burst a world and separate its fragments, it is readily seen that the fragments of largest mass would move in orbits more nearly coincident with that of the original planet, while the smaller fragments would revolve in orbits greatly inclined to the primitive one. This condition is wonderfully fulfilled among the asteroids. The larger planets, Ceres and Vesta, revolve in orbits with small inclinations to the ecliptic, while the smaller objects are in some instances found to move in planes with very great inclinations. The force necessary to burst a planet, and to give to its fragments certain orbits, has been computed by Lagrange, and he finds that in case any fragment is projected with an initial velocity one hundred and twenty-one times greater than that of a cannon ball, it would become a *direct* comet, with a parabolic orbit, while a primitive velocity one hundred and fifty-six times greater than that of a cannon ball would cause the fragments to revolve with a *retrograde* motion in the curve of a parabola. Any less powerful force would cause the fragment to revolve in ellipses; and it is probable that the force which operated to produce the asteroids was not more than twenty or thirty times greater than that of a cannon ball. Although the theory of Olbers has received new accessions of strength from the discovery of every new asteroid, it would be wrong to regard it as one of the demonstrated truths of astronomy. In the mean time, powerful efforts are making to sour the heavens, and a method of observation has been proffered to the Academy of sciences, of Paris, by all the visible fragments may be discovered within a period of four years. Should this plan, which contemplates a division of the heavens among different astronomers, be adopted, volunteers have already presented themselves, and the most interesting results may be anticipated.

From this curious branch of astronomical inquiry we turn to one of still deeper interest. In the examinations for new planets, thus far, the telescope has been the sole instrument of research. Conjectures based upon analogical reasoning, it is true, guided the instrumental examinations, but the mind had never dared to rise to the effort of reasoning its way analytically to to exact position of an unknown body. It has been reserved for our own day to produce the most remarkable and the boldest theorizing which has ever marked the career of astronomical science.—I refer to the analytic effort to trace out the orbit, define the distance, and weigh the mass of an unknown planet as far beyond the extremest known planet as it is from the sun.

I am fully aware of the difficulties by which I am surrounded, when I invite your attention to this complex and intricate subject; and I know how utterly impossible it i., in a popular effort, to do any kind of justice to the intricate and involved reasoning of the great geometers, who have not only rendered themselves, but the age in which we live, illustrious by their efforts to resolve this, the grandest problem which has ever been presented for human genius.—Trusting to your close attention, I shall attempt to exhibit some faint outline of the train of reasoning and the kind of research employed in rescuing an unknown world from the viewless regions of space in which it has been tracing its unknown orbit for ages commensurate with the existence of the great system of orbs of which it constitutes a part.

After the discovery of the planet Uranus, by Sir William Herschel, geometers were not long in fitting it with an orbit which represented in the outset, with accuracy, its early movements. With this orbit it became possible to trace its career backwards, and to define its position among the fixed stars for fifty or one hundred years previous to the date of its discovery. This was actually done, with the hope of finding that the place of the planet had been observed and recorded by some astronomer, who ranked it among the fixed stars. This hope was not disappointed. The planet, believed to be a fixed star, had been seen and observed no less than nineteen different times, by four different observers, through a period running back nearly one hundred years previous to the discovery of its planetary character by Herschel. These remote observations were of the greatest value as data for the determination of the elements of its elliptic orbit, and for the computation of the mean places, which might serve to predict its position in coming years.

A distinguished astronomer, M. Bouvard, of the Paris Academy of Sciences, about thirty years ago, undertook the analytic investigation of the movements of Uranus, and a computation of exact tables. He was met, however, by difficulties which, in the state of knowledge as it then existed, with reference to this planet, were absolutely insurmountable. He found it quite impossible to obtain any orbit which would pass through the places of the planet determined after its discovery, and through those positions which had been fixed previous to that epoch. In this dilemma it became necessary to reject the old observations as less reliable than the new ones, and the learned computer leaves the problem for posterity to resolve, carefully abstaining from any absolute decision in the case.

His orbit, based upon the new or modern observations, and his tables being computed, it was hoped that the theoretic places of the planet would thereafter coincide with the observed places and that all discrepancies which might not be fairly chargeable to errors of observation, would be removed. In this expectation, however, the astronomical world was disappointed; and while the tables of Bouvard failed absolutely to represent the ancient observations, in a few years they were but little more truthful in giving the positions actually filled by the planet under the telescope. The discrepancies between the theoretic and actual places of the planet began to attract attention many years since. As early as 1838, Mr. Airy, Astronomer Royal of England, on a comparison of his own observations with the tables, found that the planet was out of its computed track, by a distance as great as the moon's distance from the earth, and that it was actually describing an orbit greater than that pointed out by theory. It seemed that this remote body was breaking away from the sun's control, or that it was operated on by some unknown body deep sunk in space, and which thus far had escaped the scrutinizing gaze of man.

These deviations became so palpable as to attract general attention, and various conjectures were made with reference to their probable cause. Some were disposed to regard the law of gravitation as somewhat relaxed in its rigorous application to this remote body; others thought the deviation attributable to the action of some large comet, which might sway the planet from its course; while a third set of philosophers conjectured the existence of a large satellite revolving about Uranus, and from whose attraction the planet was caused to swerve from the computed orbit. These conjectures were not sustained by any show of reasoning, and were of no scientific value.

Such was the condition of the problem when it was undertaken by a young French astronomer, not quite unknown to fame in his own country, but comparatively at the beginning of his scientific career. The friend of Arago, Leverrier's Cometary Investigations, and more especially his researches of the motions of Mercury, had gained for him the confidence of this distinguished savant, and Arago urged on his young associate the importance of the great problem presented in the perturbations of Uranus, and induced him to abandon other investigations, and concentrate all the energies of his genius on this profound and complex investigation.

The extraordinary powers of Leverrier as a mathematical astronomer had been so successfully displayed in his researches of the motions of Mercury, that it deserves a passing notice. The old tables of this planet,

Leverrier believed to be defective. He therefore set about a thorough examination of its entire theory, and after a rigid scrutiny, deduced a new set of tables, from which the places of the planet might be predicted with greater precision.

The transit of Mercury across the sun's disc, which occurred on the 8th day of May, 1845, presented an admirable opportunity to test the truth of the new theory of the young astronomer. Most unhappily for his hopes, all observations in Paris were rendered impossible by the clouds, which covered the heavens during the entire day on which the transit took place. While the computer was sadly disappointed, I was more fortunate, for a pure and transparent atmosphere favored this, the first astronomical observation I ever made. A slight reference to this occurrence may be pardoned. For three years I had been toiling to complete a most difficult and laborious enterprise, the erection of an astronomical observatory of the first class, in a country where none had ever existed. Amid difficulties and perplexities which none can ever know, the work had moved on, and at length I had the high satisfaction of seeing mounted one of the largest and most perfect instruments in the world. I had arranged and adjusted its complex machinery—had computed the exact point on the sun's disc where the planet ought to make its first contact—had determined the instant of contact by the old tables, and by the new ones of Leverrier, and with feelings which must be experienced to be realized, five minutes before the computed time of contact, I took my post at the telescope to watch the coming of the expected planet. After waiting what seemed almost an age, I called to my friend how much time was yet to pass, and found but one single minute out of five had rolled heavily away. The watch was again resumed. Long and patiently did I hold my place, but again was forced to call out, how speeds the time? and was answered that there was yet wanting *two minutes* of the computed time of contact.—With steadfast eye, and a throbbing heart, the vigil was resumed, and after waiting what seemed an age, I caught the dark break which the black body of the planet made on the bright disc of the sun. Now! I exclaimed; and within *sixteen seconds* of the computed time did the planet touch the solar disc, at the precise point at which theory had indicated the first contact would occur.

The planet was followed across the disc of the sun, round and sharp, and black, and every observation confirmed the superior accuracy of the new tables of Leverrier. While the old tables were out fully a minute and a half in the various contacts, those of Leverrier were in error by only about sixteen seconds as a mean.

The great success of this investigation encouraged the young astronomer to accept the difficult task which Arago proposed for his accomplishment, and he earnestly set about preparing the way for a full discussion of the grand problem of the perturbations of Uranus. The importance of the subject demanded the greatest caution, and having determined to rely solely on his own efforts, he at once rejected all that had been previously done, and commenced the problem at the very beginning. New analytic theories were formed; elaborate investigations of the planets Jupiter and Saturn, as disturbing bodies, were made, and an entire clearing up of all possible causes of disturbance in the known bodies of the system was laboriously and successfully accomplished, and the indefatigable mathematician finally reached a point where he could say, here are residual perturbations which are not to be accounted for by any known existing body, and their explanation is to be sought beyond the present ascertained limits of the solar system.

As early as the 10th of November, 1845, M. Leverrier presented a memoir to the Royal Academy of Sciences in Paris, in which he determined the exact perturbations of Jupiter and Saturn on Uranus. This was followed by a memoir, read before the academy on the 1st of June, 1846, in which he demonstrates that it is impossible to render an exact account of the perturbations of Uranus in any other way than by admitting the existence of a *new planet* exterior to the orbit of Uranus, and whose heliocentric longitude he fixes at 325° on the 1st of January, 1847. On the 30th of August, 1846, a third memoir was presented to the academy, in which the elements of the orbit of the unknown planet are fixed, together with its mass and and actual position, with greater accuracy, giving on the 1st of January, 1847, 326° 32' for its heliocentric longitude. Finally, on the 5th of October, 1847, a fourth memoir was read, relative to the determination of the plane of the orbit of the constructive planet.

It is quite impossible to convey, in popular form, the least idea of the profound analytic reasoning employed by M. Leverrier in this wonderful investigation. None but the rarest genius would have dared to reach out 1,800,000,000 of miles into unknown regions of space, to *feel* for a planet which had displaced Uranus by an amount only about equal to four times the apparent diameter of the planet Jupiter, as seen with the naked eye— a quantity so small that no eye, however keen and piercing, without telescopic aid, could ever have detected it. Yet from this minute basis was the magnificent superstructure to be reared which should eventually direct the telescope to the place of a new and distant world. To many minds, the resolution of such a problem may appear utterly beyond the powers of human genius, and without one ray of light to illumine the midnight darkness which surrounds it to them, they are disposed to reject the entire subject. An attentive examination of the following train of reasoning may at least demonstrate that the problem is not quite so hopeless as it would at first appear.

It was not necessary to extend researches to all quarters of the heavens indifferently, in an effort to find the unknown body. All the planets revolve in planes nearly coincident with the plane of the earth's orbit, and more especially do the distant ones. Jupiter and Saturn and Uranus revolve in orbits but little inclined to the plane of the ecliptic.—Hence, it was fair to conjecture that the new planet, should it ever be found, would not violate this general law, and a search for it was properly limited to a narrow belt near the plane of the earth's orbit.—The limits of research were thus brought down to a narrow zone, sweeping around the entire heavens indeed, but insignificant in extent, when compared with the whole celestial sphere.

The next point of examination was the probable distance of the unknown planet. Here, again, analogy came to the aid of Leverrier. The empirical law of Bode, already explained in a former lecture, showed that the remote planets increase their distances by a very simple law. Saturn was twice as remote as Jupiter; Uranus was at double the distance of Saturn, and it was fair to conclude that the unknown planet would be about twice as far from the sun as Uranus. As a first approximation, then, its distance was fixed at about 3,600,000,000 of miles from the sun. Kepler's law, regulating the ratio between the distances and periods of the planets, gave at once the time of revolution of the new planet, in case its distance had been correctly assumed. In the next place, it was fair to conclude that the orbit of the new planet, like those of Jupiter, Saturn, and Uranus, would not differ greatly from a circle. These conjectures were, in some degree, confirmed by a very simple train of reasoning with reference to the distance of the disturbing body. If it revolved in an orbit very near to that of Uranus, then its effect on Uranus ought to be excessive, when compared with its influence on Saturn, which was found not to be the case. Again, if it revolved in an orbit very far beyond the limit assigned above, its effect on Uranus and Saturn would be very nearly the same, which was not verified by examination.

Having thus roughly fixed limits for the orbit of the unknown body, the work of the mathematician now commences, having for its grand object the determination of the true places of the planet sought at some given epoch, and such an orbit as will represent the perturbations of Uranus in the most perfect manner. To exhibit, in some faint degree, the difficulty of this investigation, let us conceive that up to the 1st of January, 1800, the solar system had consisted only of the known bodies, the sun, planets, satellites, and comets. The orbits of all the planets are accurately ascertained, and their reciprocal influences computed and known. The outermost planet, Uranus, revolves in its vast orbit obedient to the great law of gravitation, acknowledging the predominant influence of the sun, and swaying more or less to the action of the nearest planets, Saturn and Jupiter. Its predicted and observed places coincide, and its movement is followed with confidence and exactitude.

With a full and perfect knowledge of the orbit of Uranus, let a new planet be created and projected in a vast orbit exterior to, and remote from the orbit of Uranus. The new body thus added to the system would in-

THE DISCOVERY OF NEW PLANETS. 121

stantly derange the motions of Uranus, causing it to recede from the sun, and increasing its mean period of revolution. In this case, the total effect of the new planet on Uranus would be perturbation, and it would not be quite impossible, even for one not skilled in the higher mathematics, to see how the action of the newly created planet on the movements of the old one might actually reveal approximately the position of the disturbing body. It is manifest that when the two planets are in conjunction, or on a right line passing through them and the sun, that at this configuration the new planet would exert its greatest power to drag the old one outward from the sun, and if it could be found at what point of its orbit the old planet actually receded to its greatest distance from the sun, in the same direction, nearly, must the disturbing body have been situated at that time. In this way, we perceive, one place of the new planet might be approximately found, and from its periodic time it would be possible to trace it backward or forward in its orbit, for the present supposed to be circular.

The problem here presented is certainly sufficiently difficult, yet its complexity is very far from being equal to that presented in nature, and with which the French geometer found himself obliged to grapple.— Although unknown, the new planet did exist, and for ages had silently pursued its unknown orbit round the sun. Its influence on Uranus had been ever active, and when the observations on Uranus were made, and its places determined, from which its elliptic elements were to be derived, these very places were in part dependent on the action of the invisible disturber, and hence a portion of its influence would be darkly concealed in the orbit of Uranus; and to divide the entire effect of the new planet on the old one correctly between the disguised portion and that producing real perturbation, was attended with the greatest difficulty, and could only be reached by adopting certain positive hypotheses. Surrounded by all these difficulties, Leverrier worked on, and with consummate art, so constructed his analytical machinery as to meet and master every difficulty; and he finally announces to the world the figure of the orbit of his imaginary planet, its distance, period of revolution, and even the mass of matter it contains.

These important communications were made to the French Academy of Sciences, on the 31st of August, 1846. On the 18th of the following month, M. Leverrier wrote to his friend M. Galle, of Berlin, requesting him to direct his telescope to that point in the heavens which his computations had revealed as the one occupied by the constructive planet. The request was readily complied with, and on the very first evening of examination, a star of the eighth magnitude was discovered, which was evidently a stranger in that region, as it was not found on an accurate map of the heavens including all stars of that magnitude. The following evening was awaited with the deepest interest, to decide, by the actual motion of the suspected star, whether indeed it was the planet so wonderfully revealed by the analysis of Leverrier. The night came on. Again was the telescope directed to the star in question, when lo! it had moved from its former place, in a direction and with a velocity almost precisely accordant with the theory of the French geometer! The triumph was perfect—the planet was actually found. The news of its discovery flew in every direction, and filled the world with astonishment and admiration.

The exceeding accuracy with which its place had been predicted, coming within less than *one degree* of its actual position, gave to M. Leverrier the highest confidence in the perfection of his analysis, and filled with astonishment the oldest and most learned astronomers. If scepticism had existed with reference to the possible solution of so complex a problem—if the theory of Leverrier had been regarded as a beautiful speculation, ingenious and plausible, but vain in its practical application—the actual discovery of the planet silenced all cavil, and put to flight every lingering doubt.

As if anything were wanting, to give a more positive character to the computations of Leverrier, it was now found that a young English mathematician, Mr. Adams, of Cambridge, had actually accomplished the resolution of precisely the same problem, and had reached results almost identical with those of the French geometer. This astonishing coincidence on the part of two computers unknown to each other, each starting from the same data, pursuing independent trains of reasoning, and arriving at the same results, confirmed, as it would seem, in the fullest manner, the accuracy of the resolution which had been obtained.

On learning that Leverrier had communicated to the Academy of Sciences, in August, 1846, his final results, I wrote immediately requesting the computed place of his planet, with such directions as would best guide me in a search which I desired to make for it with the great refracting telescope of the Cincinnati Observatory. But before my letter reached its destination, the planet had been found, and the news of its discovery soon reached the United States. It was almost impossible for me to credit the statement, and I was almost disposed to believe that the prediction of the planet's position had been mistaken for its actual discovery. With these conflicting doubts, I waited for the coming of night with a degree of anxiety and excitement which may be readily imagined. I had no star chart to guide me in my search for the planet;—I had no meridian instrument with which to detect it by its motion; but I was not without hope that the power of our great telescope might be sufficient to select, at once, the planet from among the fixed stars, by the magnitude of its *disk*.

As soon as the twilight disappeared, the instrument was directed to the point in the heavens where the planet had been found. I took my place at the *finder* or small telescope attached to the larger one, and my assistant was seated at the great instrument.

THE DISCOVERY OF NEW PLANETS. 123

On placing my eye to the finder, four stars of the eighth magnitude occupied its field. One of them was brought into the field of the large telescope, and critically examined by my assistant, and rejected. A second star was in like manner examined, and rejected. A third star, rather smaller and whiter than either of the others, was now brought to the centre of the field of the great telescope, when my assistant exclaimed, "there it is! there is the planet! with a disk as round, bright and beautiful as that of Jupiter!" There, indeed, was the planet, throwing its light back to us from the enormous distance of more than 3,000.000,000 of miles, and yet so clear and distinct, that in a few minutes, its diameter was measured, and its magnitude computed.

It is not my intention to follow, critically, the history of this wonderful discovery, yet there are some facts so remarkable that it would be wrong to pass them in silence. From the moment the planet was detected in Berlin, it has been observed by all the best instruments in the world, with a view of ascertaining how accurately theory had assigned the elements of its orbit. In consequence of its very slow motion, it became a matter of the utmost importance to obtain, if possible, some remote observation made by an astronomer who might have entered the place of the planet in his catalogue as a fixed star. Mr. Adams, of England, led the way in the computation of the elements of the orbit of the new planet, from actual observation, and was followed by many other computers, among them our countryman S. C. Walker, then of the Washington Observatory.

Having obtained an approximate orbit, Mr. Walker computed backwards the places of the new planet for more than fifty years, and then examined the late catalogues, in the hope of finding its place on some of them as a fixed star. Among recent catalogues there was no success, but in an examination of Lalande's catalogue, he found an observation on a star of the eighth magnitude, made May the 10th, 1795, which was so near the place which his computation assigned the planet at the same date, that he was led to suspect that this star might indeed prove to be the new planet. In case his conjecture were true, on turning the telescope to the place occupied by the star, it would be found *blank*, as its planetary motion would have removed it very far from the place which it occupied more than fifty years before. The experiment having been made, no star could be found, and strong evidence was thus presented that Mr. Walker had actually found an observation of the new planet, giving its position in 1795; but in consequence of the great discrepancy between the period of M. Leverrier and that which would result from a reliance on this observation of the new planet Neptune, Mr. Walker's discovery was at first received with great hesitation. A greater doubt was thrown over the matter from the fact that Lalande had marked the observation as uncertain, and it w · only by reference to the original manuscripts preserved in the Royal Observatory of Paris that the doubts could be removed.

The discovery of Mr. Walker was subsequently made by Mr. Petersen, of Altona, and the results of these astronomers reached Paris on the same day. A committee was at once appointed to examine the original manuscript of Lalande, when a most remarkable discovery was made. This astronomer had observed a star of the eighth magnitude on the evening of the 8th of May, 1795, and on the evening of the 10th, not finding the star as laid down, but observing one of the same magnitude very near the former place, he rejects the observation of the 8th of May as inacurate, and enters the observation of the 10th, marking it doubtful.

On close examination, this star proves to be the planet Neptune, and by this discovery we are placed in possession of observations which render it possible to determine the elliptic elements of the new planet with great precision. These differ so greatly from those announced by Leverrier and Adams previous to the discovery, that Prof. Pierce, of Cambridge, Mass., pronounces it impossible so to extend fairly the limits of Leverrier's analysis as to embrace the planet Neptune; and that, although its mass, as determined from the elongations of its satellite, renders it possible to account for all the perturbations of Uranus by its action, in the most surprising manner, yet, in the opinion of Prof. Pierce, it is not the planet to which geometrical analysis directed the telescope. Leverrier rejects absolutely the result reached by the American geometer, and claims Neptune to be the planet of his theory, in the strictest and most legitimate sense.

Time and observation will settle the differences of these distinguished geometers, and truth being the grand object of all research, its discovery will be hailed with equal enthusiasm by both of the disputants. In any event, the profound analytic research of Leverrier is an ever-during monument to his genius, and his name is forever associated with the most wonderful discovery that ever marked the career of astronomical science.

LECTURE VIII.

THE COMETARY WORLDS.

THE wonderful characteristics which mark the flight of comets through space; the suddenness with which they blaze forth; their exceeding velocity, and their terrific appearance; their eccentric motions, sweeping towards the sun from all regions and in all directions, have rendered these bodies objects of terror and dread in all ages of the world. While the planets pursue an undeviating course round the sun, in orbits nearly circular, and almost coincident with the plane of the earth's orbit, all revolving harmoniously in the same direction, the comets perform their revolutions in orbits of every possible eccentricity, confined to no particular plane, and moving indifferently in accordance with, or opposed to, the general motion of the planets. They come up from below the plane of the ecliptic, or plunge downwards towards the sun from above, sweep swiftly round this their great centre, and with incredible velocity wing their flight far into the fathomless regions of space, in some cases never again to reappear to human vision.

In the early ages of the world, superstition regarded these wandering fiery worlds with awe, and looked upon them as omens of pestilence and war; and indeed, even in modern times, no eye can look upon the fiery train spread out for millions of miles athwart the sky, and watch the eccentric motions of these anomalous objects, without a feeling of dread. The movements of the planets inspire confidence. They are ever visible, and true to their appointed times while the comet, erratic in its course, bursts suddenly and unannounced upon the sight, and no science can predict in the outset its uncertain track—whether it may plunge into the sun, or dash against one of the planetary systems, or even come into collision with our own earth, is equally uncertain, until after a sufficient number of observations shall have been made to render the computation of the elements of its orbit possible.

Previous to the discovery of the law of universal gravitation, comets were looked upon as anomalous bodies, of whose motions it was quite impossible to take any account. By some philosophers they were regarded as meteors kindled into a blaze in the earth's atmosphere, and when once extinguished they were lost forever. Others looked upon them as permanent bodies, revolving in orbits far above the moon, and reappearing at the end of long but certain intervals. When, however, it was discovered that, under the influence of gravitation, any revolving world might describe either of the four curves, the circle, ellipse, parabola or hyperbola, it at once became manifest that the eccentric movements of the comets might be perfectly represented by giving to them orbits of the parabolic or hyperbolic form, the sun being located in the focus of the curve. According to this theory, the comet would become visible in its approach to its perihelion, or nearest distance from the sun,—would here blaze with uncommon splendor, and in its recess to the remote parts of its orbit, would gradually fade from the sight, relaxing its speed, and performing a large proportion of its vast curve far beyond the reach of human vision.

Such was the theory of Newton, and such were the fair deductions from the great law of nature which he had revealed to the world. He awaited with deep interest the appearance of some brilliant comet, whose career he might trace, in the full confidence that observation would confirm the truth of his bold hypothesis. Fortunately, his impatience was soon gratified. In the year 1680 a most wonderful comet made its appearance, which, by its splendor and swiftness, excited the deepest interest throughout the world. It came from the regions of space immediately above the ecliptic, and plunging downwards with amazing velocity, in a direction almost perpendicular to this plane, it appeared to direct its flight in such manner that it must inevitably plunge directly into the sun. This was not, however, the case. Increasing its velocity as it approached the sun, it swept round this body with the speed of a million of miles an hour, approaching the sun to within a distance of its surface of a sixth part of the sun's radius. It then commenced its recess, throwing off a train of light which extended to the enormous distance of 100,000,000 of miles. With the swiftness of thought almost, it swept away from the sun, and was gradually lost in the distant regions of space whence it came, and has never since been seen. Such were the general characteristics of the body to whose rapid motions Newton attempted to apply the law of universal gravitation.

Its positions were marked with all the accuracy which the instruments then in use permitted, and it was found that a parabolic curve could be constructed which would embrace all the places of the comet.—The great eccentricity of its orbit, and its vast period, amounting to nearly six hundred years, gave to the comet great interest, but rendered it an unfit object for successful analytic research. The great English astronomer, Halley, had studied it with the closest care, and with a rigid application of Newton's theory, he reached results quite as satisfactory as the circumstances of the case rendered possible.

Fortunately, in 1682, another comet made its appearance, to the study of which Halley devoted himself with a zeal and success which has justly stamped his name on this remarkable body; and as our limits forbid an extensive investigation of the history and theory of comets, I propose to examine this one with that degree of detail which may convey some idea of the limits of human knowledge in this complicated department of science.

At the suggestion of Newton, Halley had searched all ancient and modern records, for the purpose of rescuing any historical details touching the appearance and aspect of comets, from the primitive ages down to his own time. On the appearance of the comet of 1682, he observed its position with great care, and with wonderful pains computed the elements of its orbit.. He found it moving in a plane but little inclined to the ecliptic, and in an ellipse of very great elongation. In its ophelion, it receded from the sun to the enormous distance of 3,400,000,000 of miles.— He discovered that the nature of its orbit was such as to warrant the belief that the comet would return at regular intervals of about seventy-five years, and recurring to his historical table of comets, he found it possible to trace it back with certainty several hundred years, and with probability even to the time of the birth of Mithridates, one hundred and thirty years before Christ. At this, its first recorded appearance, its magnitude must have been far beyond anything subsequently seen, as its splendor is said to have surpassed that of the sun.

In the years 248, 324, and 399 of the Christian era, remarkable comets are recorded to have appeared, and the equality of interval corresponds well with Halley's comet. In the year 1006, it presented a frightful aspect, exhibiting an immense curved tail in the form of a scythe. In 1456, its appearance spread consternation through all Europe, and led to most extravagant acts on the part of the reigning pontiff, who actually instituted a form of prayer against the baleful influence of the comet, and thus increased the terrors of the ignorant and superstitious. The comet appeared with certainty in 1531, and again in 1607, and from an examination of all the facts, and with full confidence in his computations, Halley ventured the bold prediction that this same comet would reappear about the close of 1758, or the beginning of 1759.

This was certainly the most extraordinary prediction ever made, and the distinguished philosopher, knowing that he could not live to witness the verification of this prophetic announcement, expresses the hope that when the comet shall return, true to his computed period, posterity will do him the justice to remember that this first prediction was made by an Englishman. In the age when these investigations were made, the theory of comets was in its infancy, and it is believed by those competent to form a just opinion, that Halley was the only man living who could have computed the orbit of his comet.

As the period approached for the verification of this extraordinary announcement, the greatest interest was manifested among astronomers, and efforts were made to predict its coming with greater accuracy, by computing the disturbing effects of the larger planets within the sphere of whose influence the comet might pass. This was a new and difficult branch of astronomical science, and it would be impossible to convey the least idea of the enormous labor which was gone through by Clairaut and Lalande, in computing the perturbations of this comet through a period of two revolutions, or one hundred and fifty years.

"During six months," says Lalande, "we calculated from morning till night, sometimes even at meals,—the consequence of which was that I contracted an illness which changed my constitution during the remainder of my life. The assistance rendered by Madame Lepaute was such, that without her we never should have dared to undertake the enormous labor, in which it was necessary to calculate the distance of each of the two planets, Jupiter and Saturn, from the comet, separately for every degree, for one hundred and fifty years.

Amid all these difficulties, the computers toiled on until finally, the period coming on rapidly for the comet's return, they were forced to neglect some minor irregularities, and Clairaut announced that the comet would be retarded one hundred days by the influence of Saturn, and five hundred and eighteen days by the action of Jupiter;—he therefore fixed its perihelion passage for the 13th of April, 1759, stating, at the same time, that the result might be inaccurate by some thirty days either way, in consequence of being pressed for time, and his having neglected certain small perturbations.

These results were presented to the Academy of Sciences on the 14th of November, 1758, and on the 25th of the following December, George Palitch, an amateur peasant astronomer, caught the first glimpse of the long expected wanderer, which, after an absence of three-quarters of a century, once more returned to crown with triumph the great English astronomer who first foretold its period, and the eminent French mathematicians who had actually computed its perihelion passage, to within nineteen days, in seventy-six years!

Here, then, was a new world added to the solar system, linked to the sun by the immutable law of gravitation; sweeping out into space to the amazing distance of 3,800,000,000 of miles; lost to the gaze of the most powerful telescope, and yet traced by the human mind through its vast and hidden career, with an accuracy and precision from which there was no escape. The very small error of nineteen days in Clairaut's computation strikes us with astonishment, when we remember the imperfect state of analytical science at that day, and the fact that two planets Uranus and Neptune, which have since been discovered, were then not even suspected to have any existence.

The magnificent display which had marked some of the early returns of Halley's comet, and which produced such consternation among all classes, the educated as well as the ignorant, were not presented during its appearance in 1758. This was owing, in part, to the unfavorable position of the earth in its orbit at the time of the comet's perihelion passage.—The vast trains of light which are sometimes seen to accompany comets, are only displayed in their approach to the sun. They attain their greatest splendor while the comet is in the act of passing its perihelion, and as it recedes into space, the tail fades away, from two causes —an actual diminution from condensation, and an apparent decrease on account of increased distance.

As the comet, when near the sun, moves with increased velocity, it occupies, in general, only a short period in passing through the limits of distance from the sun within which any train of light is developed. It may happen that at one return of the same comet, the earth may occupy a point in her orbit, during its perihelion passage, which may be very near to the comet, and thus afford an opportunity of witnessing its appearance at a short distance; while, on the next return, the earth being at a remote part of its orbit while the comet is passing the sun, it may be seen only with great difficulty, or even become quite invisible. If, therefore, astronomers were obliged to depend upon a uniform physical appearance of comets at their successive returns, to determine their identity, there would scarcely be the slightest chance of ever recognizing even a single one among the many thousands which are sweeping through the regions of space.

The interval from 1759 to 1835, when Halley's comet ought to make its next appearance, had witnessed extraordinary changes in astronomy. The methods for computing planetary perturbations had been greatly improved; the planet Uranus had been added to the system, and more accurate masses of the larger planets, especially of Saturn, had been obtained. Twenty-five years before the close of the comet's period, its return commenced to interest astronomers, and prizes were offered by two academies for the most perfect theory of this remarkable body. Baron Damoiseau and M. Pontecoulant were the successful competitors for the two prizes, although several other astronomers undertook and completed the task of computing the planetary perturbations. Although the computers differed slightly in the time of the perihelion passage, the difference was due to the imperfection of the data employed, rather than to any defect in the methods of computation.

For the expected return in 1835, not only was the perihelion passage computed, but the exact route of the comet among the fixed stars was wrought out with surprising accuracy, and the precise point towards which the telescope must be directed at the time when the comet would first attain the limits of visibility. Strange and almost incredible as it must appear, guided by these predictions, M. Dumouchel, director of the observatory of the Roman College, on the evening of the 5th of August, 1835, fixed his telescope in the position indicated by computation, and, on placing his eye to the tube, lo! the comet appeared, as a faint and almost invisible stain of light on the deep blue of the heavens.

Again did science triumph, in the most remarkable manner, and the computed orbit of the comet was followed by it with the most surprising accuracy.—The perihelion passage was predicted to within nine days of its actual occurrence, a most astonishing approximation to the truth, when it is remembered that this body, far as it penetrates into space, never, even at the remotest point of its orbit, escapes from the sensible influence of the planet Jupiter. Moreover, at that time, the new planet, Neptune, was unknown, and its influence over the comet could not be taken into account.

It is interesting to remark the confidence with which astronomers relied on Halley's comet for information relative to those bodies, which inhabited the regions of space exterior to the known limits of the solar system. It was urged by every computer that the orbit of this comet would one day come to be so perfectly known, that the perturbations due to the recognized bodies of the solar system might be computed with such precision, that the residual perturbations might be pronounced to be the effect of unknown planets or comets circulating in the distant regions of space. This conjecture has been realized, although by different means, and a planet is now added to our system which revolves in an orbit so vast as to circumscribe within its limits the entire sweep of the comet; and as the orbits of Neptune and Halley's comet are inclined under an angle only of $15°$ or $16°$, a time will come when the perturbations experienced by the comet when at its aphelion, from Neptune's influence, may be so great that, but for the fortunate discovery of the cause, would have falsified, in the most unaccountable manner, the predictions of the comet's return by future astronomers. During its late return, the finest telescopes in the world were employed in a critical examination of the physical condition of Halley's comet. Elaborate drawings have been made by M. Struve, the distinguished director of the Imperial Observatory at Pulkova, Russia, with the grand refracting telescope under his charge, and also by Sir John Herschel, at the Cape of Good Hope, with a twenty feet reflector of superior power. To these beautiful drawings reference will be made hereafter.

The most wonderful changes in the magnitude and figure of the comet were observed to take place from night to night, and almost perceptible from hour to hour under the eye of the observer. The nucleus was sometimes seen sharp and strongly condensed, with more or less nebulous light around it. Sometimes a luminous crescent became distinctly visible near the nucleus, giving to the comet a most extraordinary appearance. At one time M. Struve saw the comet attended by two delicately shaped appendages of light of a most graceful form, the one preceding, the other following the nucleus of the comet. At other times it was seen to be surrounded by a sort of semi-circular veil, which, extending backwards, was lost in a double train of light, which flung itself out to a vast distance from the body of the comet.

Leaving, for the present, the consideration of the physical constitution of these eccentric bodies, we proceed to the examination of a remarkable object which bears the name of Encke's comet, in consequence of the discovery by this learned astronomer that its orbit was elliptical, and its period of revolution so short as to fall fairly within the limits of perpetual examination.

In 1818, a comet was discovered by Pons, not at all remarkable for its magnitude, for it was even invisible to the naked eye, but when the attempt was made to represent its places by a parabolic orbit, which had thus far been invariably applied to the comets, it was found impossible to assign any elongated orbit which would embrace the observed positions of the comet. After a very elaborate investigation, Professor Encke at length reached the conclusion that the orbit was not a parabola, but an ellipse of comparatively small dimensions, and that this comet was actually revolving around the sun in a period of about three years. This discovery excited a great deal of interest, for it was the first in which a short period had been detected, and efforts were at once made to identify the new member of the solar system in its preceding revolutions. Olbers determined its identity with a comet which appeared in 1795, and subsequently ascertained that another, which had been observed but twice in 1786, and from which observations no elements could be computed could be no other than the new comet of that period. In this way, observations on this interesting object were obtained, stretching through some thirty-three years, or about ten of its revolutions. This extended series of observations furnished the data for a critical examination of the elements of the comet's orbit, and Prof Encke, having discussed them with elaborate care, reached the astonishing conclusion that the magnitude of the orbit was gradually diminishing, the periodic time growing shorter from revolution to revolution, and that the comet was certainly falling nearer and nearer to the sun !

To account for this extraordinary phenomenon, the learned astronomer, having exhausted all causes known to exist in the solar system, finally, with much hesitation, announced the theory of the comet's motion in a *resisting medium*. The existence of such a medium was in direct opposition to all the received doctrines of astronomy, and the absolute necessity for its use in this instance was looked upon by astronomers with feelings of strong distrust. But Encke argued that such a medium might exist, of such exceeding tenuity as not sensibly to affect the movements of the ponderous planets, while a filmy mass of vapor, such as this comet undoubtedly was, might be very sensibly retarded in its original velocity, which would diminish continually the centrifugal force, and give to the central attraction of the sun a constantly increasing power, which would produce precisely the phenomena exhibited by the comet.

With these views, Encke predicted the reappearance of the comet in 1822. In consequence of its great southern declination at that period, it escaped all the European observers, and was only seen at Paramatta, by Rumker. The approach to the sun was in some degree confirmed by these observations, but it was impossible to reconcile all the observations with the hypothesis of a medium of given density.—The return in 1825 was not favorable for deciding the question, which had now become one of the deepest interest.

Its reappearance in 1828-9 was awaited with great anxiety by the friends and opponents of the new theory. The comet came round, and passed its perihelion approximately in accordance with the predictions, but the discrepancies from 1819 up to 1829, with any theory, were so great, as to give much perplexity to those engaged in the computations. After long and patient examination, the cause of this difficulty was finally detected. The plane of the comet's orbit makes but a small angle with the orbit of Jupiter, and when the comet is in aphelion, or farthest from the sun, it always approaches very near to the path described by the planet.

A time may then come when Jupiter shall be in the act of passing that part of its orbit very near to the aphelion point of the cometary ellipse, while the comet occupies its aphelion, bringing these bodies into close proximity, and producing excessive perturbations in the movements of this almost spiritual mass. Such, indeed, was the configuration between the returns of 1819 and 1829, on which occasion the comet was delayed in its return to its aphelion nearly nine days, by the powerful attraction of Jupiter. Under these circumstances, any error in the assumed mass of the planet would exhibit itself in an exaggerated form in the perturbations of the comet. But it was believed in the outset of this investigation, that the mass of Jupiter, employed by Laplace in his theory of the planets, and computed by Bouvard could be relied on as accurate. Indeed, Laplace had applied the calculus of probabilities, and had found that there was but one chance out of eleven millions that the mass he had adopted could be in error by the one hundredth part of its value.

Suspicion, however, having been aroused with reference to the mass of Jupiter, efforts were at once commenced to sift thoroughly the matter, and three different computers of high reputation undertook the determination of Jupiter's mass by different processes. Encke obtained a mass from the perturbations of the small planet Vesta, Nicolai from the perturbations of Juno, and Airy reëxamined the original measures of the elongations of Jupiter's satellite, made new measures, and thus obtained new data for the resolution of the problem of Jupiter's mass. The results obtained by the three astronomers agreed in a most remarkable manner, and proved incontestably that Laplace's value of the mass of this planet was in error more than *four* times the hundredth part of its value and that, instead of requiring 1,070 globes of the magnitude of Jupiter to balance the sun, only, 1,049 were necessary.

With the new mass of Jupiter it seemed possible by admitting a resisting medium, to account for all the perturbations of Encke's comet, and for a time this theory seemed to receive greater consideration from distinguished men. The appearance of Halley's comet in 1835 again threw great doubt over the subject, for it was found impossible to reconcile the movements of the two comets with any assumed density of a resisting medium. Some have been disposed to adopt the idea that the revolution of the planets, for ages, in the same direction, in this supposed ethereal fluid, has impressed upon it a certain amount of motion in the same direction, and that those comets which chance to revolve with the current will be found to be operated upon differently from those which may happen to come into our system in a direction opposed to the current.

I confess, frankly, that my own mind has always revolted against the doctrine of a resisting fluid.—There are so many ways in which the single phenomenon of the gradual approach of Encke's comet to the sun may be accounted for, without resorting to an hypothesis which involves the entire destruction of the planetary system, whose perpetuity has been so effectually provided for by the great Architect of the universe, that it would require the most unequivocal testimony to secure the full consent of my own mind to the adoption of this remarkable theory. It is proper, however, to say, that it has long been received with favor by men to whose judgment I am generally disposed to yield with implicit confidence.

Leaving the further consideration of this subject for the present, we proceed to the examination of another comet of short period, which has excited great attention, especially in its recent return. As early as 1805, Prof. Gauss, in computing the elements of the orbits of the comets of that year, found one which seemed to complete its revolution in about six years. This comet, however, was lost sight of, and it was not until 1826 that M. Biela discovered the same comet on its return to its perihelion. This discovery appears to have been the result of computation, but how far the investigation was carried, I have never been able to learn.

The same object was also discovered by M. Gambart about the same time, who, on fixing its elements, found that it performed its revolution about the sun in an ellipse, with a period of six and three-quarter years. This comet, like Encke's, is only to be seen with the telescope. It presents no solid, or even well defined nucleus, and appears to be a mere vapory mass, of exceeding tenuity. Taking into account the disturbing influence of Jupiter, the returns of Biela's comet, as predicted, agreed well with observation, and gave confidence in the theory on which the predictions were founded.

The return in 1832 excited the liveliest interest throughout the civilized world, in consequence of the fact that it was discovered from computation, that on the night of the 29th of October, this comet would pass a little within the earth's orbit, and those unacquainted with the subject received the impression from this announcement, that the earth and comet would come into collision, producing the most terrific consequences. Such was the consternation excited, throughout the city of Paris especially, that the Academy of Sciences found it necessary to give to the subject their serious attention, and finally gave the matter in charge to M. Arago, who produced an elaborate report on the subject of comets generally, which served to calm the popular apprehension, and has proved to be a valuable addition to our knowledge on this difficult subject.

In this report, M. Arago, showed that the comet would indeed cross the earth's track at the time predicted, but at the moment of crossing, the earth would be some fifty-five millions of miles distant from the point occupied by the comet, and could not experience the slightest possible influence from such a body, at such a distance.

If the comet had been delayed in its approach for thirty days, by any disturbing cause, then, indeed, the earth and comet would have filled at the same time the point where their orbits intersect, and the dreaded collision would have taken place. The consequences of such a shock it is impossible to conjecture, but reasoning from the known physical condition of the comet, none of the terrible disasters so generally anticipated would have occurred. The exceeding rarity of the matter composing this body may be inferred from the statements of Sir John Herschel. "It passed," says he, "over a small cluster of most minute stars of the 16th and 17th magnitude; and when on the cluster presented the appearance of a nebula resolvable, and partly resolved, the stars of the cluster being visible through the comet. A more striking proof could not have been offered of the extreme translucency of the matter of which the comet consists. The most trifling fog would have effaced this group of stars, yet they continued visible through a thickness of cometic matter which, calculating on its distance and apparent diameter, must have exceeded 50,000 miles, at least towards its central parts. That any star of the cluster was *centrally* covered, is indeed more than I can assert; but the general bulk of the comet might be said to have passed centrally *over the group.*"

Such is the nature of the body from whose contact the ignorant apprehended the most fearful convulsions. Olbers, who studied the subject with great care, was disposed to think that in case the earth had passed directly through the comet, no inconvenience would have occurred, and no change beyond a slight influence on the climate would have been experienced.

It is useless to speculate with reference to the probable consequences of a collision, which there is scarcely one chance in millions can ever occur. Science has as yet discovered no guarantee for any planet against the possible shock of a comet; but an examination of the delicate adjustments of our own system, and those of Jupiter and Saturn, would seem to indicate to us that in all past time no derangement has ever occurred from such a cause.

The last return of Biela's comet was marked by a phenomenon unexampled, so far as I know, in the history of these wandering bodies. True to the predictions of Santini, the comet first became visible on the evening of the 26th of November, 1845, and in the precise point which had been assigned by theory. De Vico, the director of the observatory at Rome, was the first to catch a glimpse of the expected comet. Nothing remarkable in its appearance was noticed, until about the 29th of December, when Mr. E. C. Herrick, of New Haven, pointed out to several friends what he regarded to be a sort of anomalous tail, but shooting out from the head of the comet in a direction entirely at variance with the usually received theory, that the tail is always opposite to the sun. In this supposed tail a kind of *knot* was noticed, brighter and more condensed than any other part. Owing to insufficient optical power, the true character of the phenomenon was not fairly detected by Mr. Herrick.

On the night of the 12th of January, 1846, Lieut. Maury, in charge of the observatory at Washington City, discovered that what had hitherto appeared as a single body, was actually composed of *two distinct and separate comets*. This most extraordinary fact was immediately announced, and the double character was observed at all the principal observatories in Europe and the United States. There can be no doubt whatever as to the reality of the appearance. The comet actually became double, and the two parts, bound together by some inscrutable bond, continued their swift journey through space, pursuing almost exactly the route predicted for the single comet.

From measures obtained by Prof. Challis, of Cambridge, England, on the 23d of January, 1846, the two comets were separated from each other by a distance equal to about one-thirteenth the apparent diameter of the sun. On the 28th of the same month, Sir John Herschel records the following notices:—" The comet was evidently double, consisting of two distinct nebulæ, a larger and a smaller one, both round, or nearly so,— the one in advance faint and small, and not much brighter in the middle; the one which followed nearly three times as bright, and one and a half times larger in diameter, and a good deal brighter in the middle, with an approach to a stellar point.

On the evening of the 9th of February, having returned to the observatory at Cincinnati, after an absence of more than two months, I had an opportunity of beholding, for the first time, these wonderful objects, with the twelve inch refractor. The moon was nearly full, and yet the comets were distinctly visible, both included within the limits of the field of view of the instrument, and separated from each other by a distance equal to about the eighth part of the sun's diameter. The preceding comet was evidently the brighter of the two.

Clouds prevented a continuous examination of the comets from night to night, but on the evening of the 21st of February, I was surprised to find a remarkable change in the relative brilliancy of the two parts. On that evening the following comet was very decidedly brighter than its companion, and from observations made elsewhere, the change of relative brightness seems to have been effected about the 13th or 14th of February. The change was observed by Prof. Encke, of Berlin, as early as the 14th. On the evening of the 21st of February, both comets exhibited distinct trains of light, extending from the sun, and in directions parallel to each other. The centre of the nucleus of each comet was brighter than the surrounding portions, but there was no stellar point visible. The nebulosity of the two points did not intermingle.

The distance between the comets increased from day to day, until, on the 25th of February, they were separated by an amount equal to 445 seconds of arc, or between a fourth and fifth part of the sun's diameter. A part of the increase of distance was only apparent, arising from the approach of the comets to the earth, but the comets were actually receding from each other while pursuing their rapid flight through space.

Neither did the line joining the central points of the comets remain parallel to itself. From the 23rd of January to the 11th of February, this line shifted its position by an amount of angular motion equal to 8° as is shown by a comparison of the measures of Challis and Encke. By the 21st of February, this angular motion had been nearly destroyed by a retrograde movement, and thus the comets were seen to oscillate about each other, according to some mysterious law which has never been revealed. Such is a brief sketch of the phenomena presented by Biela's comet in its late return. Its next appearance will be looked for with deep interest, to confirm or destroy certain theories which have been propounded to explain its duplex character.

While the periods of the comets which we have thus far considered are comparatively short, those of others which have visited our system have been ascertained to extend to many thousands of years. The great comet of 1811, one of the most brilliant of modern times, in consequence of its remaining visible for nearly ten months, gave ample opportunity for the investigation of the elements of its orbit. After a careful investigation, M. Argelander fixes its period of revolution at 2,888 years. Bessel had examined the same subject previously, and probably with less attention, but obtained a period even greater than Argelander's, amounting to 3,383 years.

The comet of 1807 also occupied the attention of Bessel. A long series of observations furnished the data for computing its elements. The periodic time was fixed at 1,543 years. These computations are necessarily only approximate. The difficulty of obtaining accurately the periodic time increases with the length of the period, and all that can be done is to fix a limit below which it cannot fall. These vast periods give to us the means of learning somewhat of the great distance to which these objects penetrate into space. The comet of 1811, having a period probably three thousand times greater than that of our earth, must revolve at a mean distance from the sun of more than 80,000,000,000 of miles, and in consequence of its very near approach to the sun at its perihelion, its greatest distance cannot fall below 160,000,000,000 of miles!

Great as this distance is, it is perfectly certain that there are many comets which revolve in orbits far more extensive than the one described by the comet of 1811. Indeed, there seems to be no limit to the distance to which these bodies may sweep outward from the sun; and their return depends simply on the fact whether they recede so far as to fall within the attractive influence of some other sun, towards which they begin to urge their flight, and through whose system of planets they carry the same apprehensions of danger which have been caused in our own.

In reflecting on these singular objects we are led to inquire what they are, whence their origin, and by what laws are the vast trains of light which occasionally distinguish them developed? Arago divides comets into three classes, with reference to their physical constitution. He thinks they occasionally appear round, and with well defined planetary disks, showing them to be solid opaque bodies, in all respects resembling planets, and only differing from these in the great eccentricity of their orbits. In confirmation of this opinion, he asserts that comets have been seen to transit the sun, and when passing between this luminary and the eye of the spectator, they appear round and black, like the planets Mercury and Venus, when seen under the same circumstances. An example of this kind occurred on the 18th of November, 1826, when the transit of a comet across the sun was witnessed by two persons, widely separated from each other.

A second class of comets comprehends those in which there is a nucleus, but devoid of opacity, permitting the light to penetrate through even that portion which may possibly be solid. The third class, and that by far the most numerous, comprehends those comets destitute, entirely, of any solid nucleus, consisting of matter so attenuated as to compare fairly with nothing of which we have any knowledge on the earth's surface. The comets named for Encke and Biela appear to belong to this class; and even Halley's comet, according to the opinion of Sir John Herschel, seems, at its last return, to have been entirely turned into vapor in its perihelion passage.

No theory, with the exception Laplace's nebular hypothesis, has ever been framed to explain the origin of these wandering bodies. This is not the place to enter into a full development of this subject.—A few hints only can be given. Laplace, following up the speculation of Sir William Herschel, applied the theory of that astronomer to the formation of the solar system, comprehending the comets, as well as the planets and their satellites. This theory supposes that the original chaotic condition of the matter of all suns and worlds was nebulous, like the matter composing the tails of comets. Under the laws of gravitation, this nebulous fluid, scattered throughout all space, commences to condense towards certain centres. The particles moving towards these central points, not meeting with equal velocities, and in opposite directions, a motion of rotation is generated in the entire fluid mass, which in figure, approximates the spherical form.

The spherical figure once formed, and rotation commenced, it is not difficult to conceive how a system of planets might be produced from this rotating mass, corresponding, in nearly all respects, to the characteristics which distinguish the planets belonging to our own system. If, by radiation of heat, this nebulous mass should gradually contract in size, then a well known law of rotating bodies would insure an increased velocity of rotation. This might continue until the centrifugal force, which increases rapidly with the velocity of the revolving body, would finally come to be superior to the force of gravity at the equator, and from this region a belt of nebulous fluid would thus be detached in the form of a ring, which would be left in space by the shrinking away of the central globe. The ring thus left would generally coalesce into a globular form, and thus would present a planet with an orbit nearly, if not quite circular, lying in a plane nearly coincident with the plane of the equator of the central body, and revolving in its orbit in the same direction in which the central globe rotates on its axis.

As the globe gradually contracts, its velocity of rotation continually increasing, another ring of matter may be thrown off, and another planet formed, and so on, until the cohesion of the particles of the central mass may finally be able to resist any further change, and the process ceases.

The planetary masses, while in the act of cooling and condensing, may produce satellites in the same manner, and by the operation of the same laws by which they were themselves formed. Strange and fanciful as this speculation may appear, there are many facts which tend strongly to give it more than probability. It accounts for all the great features of the solar system, which, in its organization, presents the most indubitable evidence that it has resulted from the operation of some great law.

The sun rotates on an axis in the same direction in which the planets revolve in their orbits; the planets all rotate on their axes in the same direction; they all circulate around the sun in orbits nearly circular, in the same direction, and in planes nearly coincident with the plane of the sun's equator. The satellites of all the planets, with one single exception revolve in orbits nearly circular, but little inclined to the equators of their primaries, and in the same direction as the planets. So far as their rotation on axes has been ascertained, they follow the general law.—In one instance alone we find the rings of matter have solidified in cooling, without breaking up or becoming globular bodies. This is found in the rings of Saturn, which present the very characteristics which would flow from their formation according to the preceding theory. They are flat and thin, and revolve on an axis nearly, if not exactly, coincident with the axis of their planet. Their stability, as we have seen, is guaranteed by conditions of wonderful complexity and delicacy, and the adjustment of the rings to the planet, (humanly speaking), would seem to be impossible after the formation of the planet.—At least it is beyond our power to conceive how this could be accomplished by any law of which we have any knowledge, and we must refer their structure at once to the fiat of Omnipotence.

Granting the formation of a single sun by the nebular theory, and we account at once for the formation of all other suns and systems throughout all space; and according to the advocates of this theory, the comets have their origin in masses of nebulous matter occupying positions intermediate between two or more great centres, and held nearly in equilibrio, until, finally, the attraction of some one centre predominates, and this uncondensed filmy mass commences slowly to descend towards its controlling orb. This theory would seem to be sustained, so far as a single truth can sustain any theory, by the fact that the comets come into our system from all possible directions, and pursue their courses around the sun either in accordance with, or opposed to, the direction in which the planets circulate. Their uncondensed or nebulous condition results from the feeble central attraction which must necessarily exist in bodies composed of such small quantities of matter. Moreover, in some cases at least, there is reason to believe, that in their perihelion passage they are entirely dissipated into vapor by the power of the sun's heat, and may thus revolve for ages, going through alternate changes of solidification and evaporation.

But whence come the enormous trains of light which sometimes attend these wandering bodies?—The last return of Halley's comet has furnished the data for the positive illustration of this mysterious subject. Sir John Herschel, after a careful and most elaborate examination of all the physical characteristics of this comet, comes to the conclusion that the figure of the comet, envelope and tail, could not be a figure of equilibrium under the law of gratification. He is therefore compelled to bring in a *repulsive force* to explain the phenomena.

I cannot do better than to quote his own language in this bold introduction of a new power.—" Nor let any one be startled at the assumption of such a repulsive force as here supposed. Let it be borne in mind that we are dealing (in the tails of comets) with phenomena utterly incompatible with our ordinary notions of gravitating matter. If they be material in that ordinary received sense which assigns to them only inertia and attractive gravitation, where, I would ask, is the force which can carry them round in the perihelion passage of the nucleus, in a direction pointing continually *from* the sun, in the manner of a rigid rod, swept round by some strong directive power, and in contravention to all the laws of planetary motion, which would require a slower angular motion of the more remote particles, such as no attraction to the nucleus could give them, be it ever so intense? The tail of the comet of 1680, in five days after its perihelion passage, extended far beyond the earth's orbit, having, in that brief interval, shifted its angular position nearly 150°. Where can we find, in its gravitation either to the sun, or to the nucleus, any cause for this extravagant sweep?

"But again, where are we to look (if only gravity be admitted) for any reasonable account of its projection *outward from the sun*, putting its angular motion out of the question? Newton calculated that the matter composing its upper extremity quitted the nucleus only two days previous to its arriving at this enormous distance."

Herschel argues the inadequacy of gravitation to account for these wonderful phenomena. The velocity with which the matter composing the tail shot forth from the head of the comet, *from the sun*, was far greater than that which the sun could impress on a body falling to it, even from an infinite distance.—An energy of a different kind from gravitation, and far more powerful, must exist, to produce such results. If, then, we are forced to the admission that a power exists in the sun capable of repelling matter of a certain quality existing in comets, a way is opened for the explanation of some of the most difficult problems with which the mind has been obliged to contend.

The diminution of the periodic time of Encke's comet has led some astronomers to adopt the idea of the existence of a resisting medium. But in case the sun possesses the power of repelling the matter of comets in their perihelion passage, a part of the matter thus repelled may be driven entirely beyond the attractive influence of the nucleus, and be irrecoverably lost. In this case, a diminution of mass would inevitably involve a like diminution of periodic time, a contraction of the orbit, and all the phenomena presented by this mysterious object. Herschel even thinks it possible, on this theory, to account for the separation of Biela's comet into two distinct objects, and it appears to me that it presents the most reasonable explanation of the luminous appearance seen at certain seasons of the year, called the *zodiacal light*.—This phenomenon appears to be a ring of nebulous matter surrounding the sun, and some of whose particles are sustained at a much greater distance than could be accounted for by gravitation. Admitting the repulsive power already adverted to, there is no difficulty in understanding how this nebulous ring may be sustained at a vast distance from the sun.

Here we freely admit that we enter the confines of the unknown. We have left the solid ground of truth and certainty and are pushing our investigations into the dim twilight of the invisible and uncertain. But as antiquity predicted that the time would come when the comets would be traced in their career, their periods revealed, and their orbits ascertained, so we may confidently hope that, at no very distant day, all the mysteries which hang around these chaotic worlds will be fully revealed, and a knowledge of their physical condition shall reward the long study and deep research of the human mind.

LECTURE IX.

THE SCALE ON WHICH THE UNIVERSE IS BUILT.

THUS far our attention has been directed to an examination of the achievements of the human mind within the limits of our own peculiar system. We have swept outward from the sun through the planetary worlds, until we have reached the frontier limits of this mighty family. Standing upon the latest found of all the planets, at a distance of more than 3,000,000,000 of miles from the sun, we are able to look backwards, and examine the worlds and systems which are all embraced within the vast circumference of Neptune's orbit. An occasional comet, overleaping this mighty boundary, and flying swiftly past us, buries itself in the great abyss of space, to return after its "long journey of a thousand years," and report to the inhabitants of earth the influences which have swayed its movements in the invisible regions whither it speeds its flight.

The magnificence and complexity of the great system of planets and satellites and comets which constitute the sun's retinue ; the immense magnitude of some of these globes; their periods of revolution, and reciprocal action, would seem to furnish a sufficient exercise, not only for the highest intellectual efforts, but for the entire energy which the human mind can exert. But the whole of this stupendous scheme, as we shall soon see, is but an infinitesimal portion of the universe of God, one unit among the unnumbered millions which fill the crowded regions of space.—Standing at the verge of the planetary system, we find ourselves surrounded by a multitude of shining orbs, some radiant with splendor, others faintly gleaming with beauty. The smallest telescopic aid suffices to increase their number in an incredible degree, while with the full power of the grand instruments now in use, the scenes presented in the starry heavens become actually so magnificent as to stun the imagination and overwhelm the reason. Worlds and systems, and schemes and clusters, and universes, rise in sublime perspective, fading away in the unfathomable regions of space, until even thought itself fails in its efforts to plunge across the gulf by which we are separated from these wonderful objects.

In our measurements within the limits of the solar system, the radius of the earth's orbit has sufficed for a unit with which to exhibit the distances of the planets and comets. Great as is this unit, measuring no less than 95,000,000 of miles we shall soon find it far too minute and insignificant to serve in our researches with reference to the grand scale of the visible universe. To obtain comprehensible ideas with reference to the interstellar spaces, we shall be obliged to call to our aid a unit, not exactly of distance, but of velocity; and before entering on the full exhibition of the main object of this lecture, permit me to direct your attention to a remarkable discovery, by which the important fact has been revealed, that light does not pass instantly from a luminous body to any remote object on which it may fall, but with a progressive motion, whose actual velocity has been ascertained. The important bearing of this discovery will become apparent as we advance in our examination of the sidereal heavens.

After the motions of the four moons of Jupiter had been sufficiently observed to construct tables of their movements, with a view to predict their eclipses, some unaccountable phenomena presented themselves, which, for a long time, baffled all efforts to explain them. It should be remembered, that the orbit of Jupiter encloses that of the earth, and when the two planets happen to be on the same side of the sun, and in a straight line passing through that orb, they are then at their least distance from each other, and are said to be in conjunction. Now suppose Jupiter to remain stationery, at the end of half a year the earth will have reached the opposite point of her orbit, and will now be more distant from Jupiter by an amount equal to the diameter of her orbit, or nearly 200,000,000 of miles. Retaining carefully these positions in the mind, we shall follow the facts about to be presented with the greatest ease.

It was found that those eclipses of Jupiter's satellites, which occurred while the earth and planet were at their least distance from each other, always came on *sooner* than the time predicted by the tables; while, on the contrary, those which took place when the planets were most remote from each other, occurred *later* than the computed time. A still more extended examination of these remarkable phenomena demonstrated the fact, that the discrepancies depended evidently on the absolute increase and decrease of distance which marked the relative position of the planets in their revolutions around the sun. For a long time, no explanation of these undeniable truths could be found, until the mystery was finally solved by Roemer, a Danish astronomer, who, with admirable sagacity, traced these irregularities to their true source, and found that they arose from the fact that light traveled through space with a finite and measurable velocity.

The explanation is simple. When Jupiter and the earth are at their least distance from each other, the stream of light flowing from the satellite of the great planet traverses a shorter space to reach the eye of the observer on the earth, by nearly 200,000,000 of miles, than when the planets are most remote from each other. In case, therefore, this stream is in any way cut off, it will run out sooner in the first than in the second position, by the time required to pass over the diameter of the earth's orbit. The stream of light is actually shorter, by 200,000,000 of miles, in the first than in the second position of the planets.

THE SCALE ON WHICH THE UNIVERSE IS BUILT.

Now the satellites of Jupiter receive their light from the sun;—they reflect this light to the earth, and when the body of their primary is interposed between them and their source of light, they are eclipsed; their light is cut off; its flow is interrupted; and when the stream of light starting from them at the instant the supply is cut off shall have run out, then, and not till then, does the satellite become invisible. This explanation accounted for all the phenomena in the most beautiful manner.

The tables had been constructed from the mean of a great number of observed eclipses. Hence, those which took place while Jupiter and the earth were near to each other, would happen earlier than prediction; while those taking place when the planets were at their greatest distance, would occur later than the time given by the tables. But the velocity with which this mysterious, subtle, intangible substance called light, flew through the regions of space, as determined by this wonderful theory, was so great as to startle the minds of even its strongest advocates, and to demand the most positive testimony to induce the belief of those disposed to scepticism. It was found to traverse a distance equal to the entire diameter of the earth's orbit, or 190,000,000 of miles, in about 16 minutes!—giving a velocity of 12,000,000 of miles per minute, or 192,000 miles in each second of time!

It is not our purpose to enter into any investigation as to the true theory of light, whether it be an actual emanation from a luminous body of material particles, or whether it be a mere vibratory or undulating motion produced by luminous bodies on some ethereal medium. My only object, at this time, is to assert the undoubted fact, that in case a luminous body were to be suddenly called into being, and located in space at the distance of 12,000,000 of miles from the eye of an observer, who was on the look-out for its light, this light would not reach him until *one minute* after the creation of the object; and should it suddenly be struck from existence, the same observer would behold it for one minute after the extinction.

Should any mind revolt from these statements—should the difficulty of the investigation, and the incredible velocity of light, demand higher and better evidence, before full faith can be given to the theory, I can only say that this evidence shall be given before we close this discussion, and with a fullness and clearness which shall set all doubts at defiance.

I now proceed to an examination of the great problem of the parallax of the fixed stars, a problem which has taxed the ingenuity of the greatest mind, and which has called into requisition the most admirable skill for a period of more than 300 years. A familiar explanation of the nature of this problem may prepare the way for a rapid sketch of the various means which have been employed in its solution. If it were possible to measure on the earth's surface a base line of a thousand miles in length, by locating an observer at each extremity of this base' with instruments suitable to fix the moon's place among the fixed stars, the telescopes of those two observers, directed to the moon's centre at the same instant would incline towards each other, and the visual ray from each of those instruments would meet at the moon's centre, and form an angle with each other. This angle, or opening of the visual rays, is called the *parallax*, and in case the object under examination were a fixed star, then would the angle in question be called the *parallax of the fixed star*.

It is readily seen that when the length of the base is known, and the parallactic angle measured, then the length of the visual ray may be at once determined, and the distance of the object is made known by the simplest rules of geometry. Parallax, then, in general, is the *apparent* change in the place of an object occasioned by the *real* change in the place of the spectator.

The whirling of the trees of a forest, produced by the rapid speed of the beholder along a railway, is a parallactic motion, and becomes less and less perceptible as the velocity of the spectator diminishes, or as the distance of the seemingly moving object becomes greater. To measure the distance of the fixed stars is then equivalent to determining the amount of parallactic change in their relative positions occasioned by the actual change of the positions from which they may be viewed by a spectator on the earth's surface.

With the sun and moon and planets, a base line equal to the earth's diameter, or about 8,000 miles has sufficed to produce a sensible and measurable parallax; but when we extend our visual rays to a fixed star, from the extremities of this base, their directions, to our senses, are absolutely parallel, or, in other language, the parallax arising from such a base is perfectly insensible. This first effort indicates, at once, the vast distance of the objects under examination; for such is the accuracy with which minute spaces are now divided, that parallax may be detected in case the object is even 160,000 times farther distant than the length of the base line.

When the orbital motion of the earth was first propounded by Copernicus, and it was asserted to revolve in an ellipse of nearly 600,000,000 of miles in circumference, and with a motion so swift that it passed over no less than 68,000 miles in every hour of time, the opponents of these startling doctrines exclaimed No! this is impossible; for if we are sweeping around the sun in this vast orbit, and with this amazing velocity, then ought the fixed stars to whirl round each other, as do the forest trees to the traveler flying swiftly by them.

But the stars of heaven do not move. Seen from any point, and at any time, their places are ever the same,—fixed, immutable, eternal,—the bright and living witnesses of the extravagance and absurdity of this new and impossible theory. To this reasoning, which was well founded, and without sophistry, the Copernicans could only reply, that such was the enormous distance of the fixed stars, that no perceptible change was occasioned by the revolution of the earth in its orbit. But this was mere assertion and the opponents met the statement by this very plain exhibition of the case.—You who believe in the doctrines of Copernicus assert that the earth revolves on an axis, which, as it sweeps round the sun, remains ever parallel to itself. This axis prolonged meets the celestial sphere in a point called the north pole. Now as the earth describes an orbit of nearly 200,000,000 of miles in diameter, its axis prolonged will cut out of the sphere of the heavens a curve of equal dimensions, and the pole will appear to revolve and successively fill every point of this celestial curve in the course of the year. Now the north pole does not revolve in any such curve; it is ever fixed, and your theory is false. The Copernican could only reply, that all the premises were true, but that the conclusion was false. The pole of the heavens did revolve in just such a curve as stated but such was the distance of the sphere of the fixed stars, that this curve of 200,000,000 of miles in diameter was reduced to an invisible point!

Three hundred years have rolled away since this controversy began. The struggle has been long and arduous. The mind, baffled in one direction, has directed its energies in another—failing in one mode of research, it devises another, and thus struggling onward for three long centuries, it at length triumphs. The facts are developed and the truth of the grand theory of Copernicus is vindicated and established, and the accuracy of these incredible statements is proved in the clearest manner.

As this discussion exhibits, clearly and beautifully, the progressive advances of human genius, I shall be pardoned for entering, at some length, into an examination of the various attempts which have been made to resolve the problem of the parallax of the fixed stars. Indeed, the distance of the nearest fixed star is to become the unit of measure with which we are to traverse the innumerable worlds and systems by which we are surrounded, and on the accuracy with which it shall be determined will depend the correctness of the survey which we are soon to make.

Failing entirely in obtaining any parallactic angle with a base line of 8,000 miles in length, the earth was employed to transport the observer from the first point of observation to a distance of 190,000,000 of miles, there again to erect his telescope, and to send up his second visual ray to the far distant star, in the hope of finding a parallactic angle with a base of such enormous extent.

Permit me to illustrate the nature of this investigation, Suppose from the centre of a plane a solid granite rock, deep sunk and immovable, rears its head far above the mists and impurities which float in the lower air. Ascending to the summit, the astronomer hews out some rough peak into the form of a vertical shaft. To this solid shaft he bolts the metallic plates which shall bear his telescope. The instrument is of a size and

STRUCTURE OF THE UNIVERSE.

power commensurate with the grand objects which it is required to accomplish. Placed in a position such that its axis shall be exactly vertical it is screw-bolted and iron-bound to the solid rock with fastenings which shall hold it from year to year, fixed and immovable as its rocky base.

To give more perfect precision to his work, the astronomer places in the focus of his eye-piece two delicate lines made from the spider's web, of a minuteness almost mathematical, which by crossing at right angles, form a point of the utmost precision exactly in the axis of the telescope. These are in like manner fixed immovably in their places, and now the machinery is prepared with which the observations are to be conducted.

Suppose the observations to commence to-night.—On placing the eye to the telescope, and looking directly up to the zenith, a star enters the field of the instrument, and borne along by the diurnal motion of the heavens, advances towards the central point determined by the intersection of the spider's lines. In passing across the field of view, its minute diameter is exactly bisected by one of these delicate lines, and the exact moment, to the hundredth part of a second of time, is noted at which it passes the central point. This observation completed with all posssble precision, in case no change in the apparent place of the star is produced by the revolution of the earth in its orbit, or by any other cause, on each successive night throughout the entire year the same phenomena will be repeated in the same precise order. When the hour comes round, the star will enter the field, thread the spider's line, and reach the central point at the same precise instant, night after night even for a thousand revolutions of the earth on its axis.

Such, then, is the delicate means employed in the examination of the problem of the parallax of the fixed stars; and nearly in this way did Bradley, the great English astronomer, prosecute this intricate investigation. If any change in the star's place is occasioned by the revolution of the earth in its orbit, sweeping, as it does, the spectator round the circumference of a track nearly 200,000,000 of miles in diameter, it is easy to compute, not the amount, but the direction in which these changes will be accomplished. These computations were made by the astronomer, and all things being prepared, he commenced the series of observations which were to lead to the most important results. The discovery of absolute fixity in the star would be a great negative result, and any changes, no matter of what kind or character, could not fail to be detected.

Night after night was the astronomor found at his post, and as the months rolled slowly away, he began to perceive that his star, which, for a long time threaded the spider's line as it was in the act of passing the field of the telescope, began slowly to work off from this line, at last absolutely separating itself from it, and failing to reach the central point of the field at the precise instant first recorded. It soon became manifest that some cause or causes were operating to produce an apparent change in the place of the star, but what was the astonishment of Bradley to find that the changes in question could not be produced by parallax, for the motions detected were almost precisely opposite to those which would arise from this cause.

Long years of laborious examination were finaly rewarded with two of the most brilliant discoveries ever accomplished by human skill and genius. The first of these demonstrated the fact that the sun and moon were so operating on the protuberant matter at the earth's equator as to cause the axis of the earth to oscillate or revolve in a minute orbit, nodding to and fro under the influence of the configurations of these two controlling bodies, and following, in the most absolute manner, their relative positions. The effect of this variation, called *nutation*, is to cause all the stars to appear alternately to approach and recede from the pole. —The real effect is to move the pole by the same amount.

The value of this change has been determined with the utmost precision, and although its entire effect does not shift the pole over a greater space than the fourth part of the apparent diameter of the planet Jupiter, its values, as deduced by different astronomers and by different processes, scarcely differ by the fraction of a second of space. As a specimen of the accuracy attained in these delicate measurements, I will give three values recently obtained by the Russian astronomers.—M. Busch, from Bradley's observations, obtains the value $9''.2320$; M. Liendhal, from observations at Derpat, finds the value $9'',1361$; M. Peters, from right ascensions of Polaris, observed at Derpat fixes the value at $9''.2164$.—The mean of the three values is $9''.2231$, the highest difference from which is less than the tenth part of one second of arc.

Valuable as was this discovery, it was actually surpassed by the importance of the second, for which we are in like manner indebted to Bradley. This second phenomenon consisted in an apparent movement of all the fixed stars in a minute orbit, which was accomplished in a year for every individual, and showed, in the most absolute manner, that it depended in some way on the orbital revolution of the earth.—For a long time, the true explanation of this phenomenon, which Bradley saw at once was not parallactic, eluded his highest sagacity. Potent thought and persevering reflection were, however, at last triumphant, and an explanation was finally reached, not only of the most satisfactory kind, but involving nothing less than an absolute demonstration of the orbital motion of the earth, and a full confirmation of the velocity of light, whose prodigious swiftness had staggered the faith of many anxious to credit so marvelous a statement.

A few words will suffice to explain these phenomena. If we admit the progressive motion of light and the revolution of the earth in its orbit, it is manifest that the celestial bodies will not occupy in the heavens the places they appear to fill. Take, for example, the planet Jupiter, and even suppose the planet to be fixed. The telescope is directed to this object, and the light from the planet, streaming through the axis of the instrument, reaches the eye of the observer, and produces the visible image of the planet. But these very particles of light have occupied nearly 40 minutes in passing from the planet to the eye of the observer. During these 40 minutes, the earth has progressed in its orbit some 37,000 miles, and the spectator on the earth, borne along with it, must see the planet, not where it actually is but where it was in appearance some forty minutes before. The same effect, in kind, is produced on the places of the fixed stars, and is called *aberration*. Understanding, now, that *some* effect must arise from these causes, (the velocity of light and the motion of the earth), let us endeavor to render its nature clear, and the results palpable. To accomplish this, we must resort to the simplest means of elucidation.

Suppose a person were on the deck of a boat floating down the current of a river at any given rate per hour. As he moves steadily down the stream, he catches sight of an object on the shore, through which he proposes to send a rifle ball. The marksman will not aim directly at the object. Why? Because he knows that the rifle ball will partake of the boat's motion, and will be carried down, after it leaves the gun and before it reaches the mark, a distance equal to the progressive motion of the boat during the time of flight of the ball. To strike the mark, he must therefore make this necessary allowance, and aim above it the required quantity. It is readily seen that the faster the boat moves, the greater will this allowance be.

Now reverse the proposition, and suppose a rifle fixed on shore, and so directed as to fire a ball *down the barrel* of a gun on a moving boat. In case the two rifles are on the same exact level, and the axes of the barrels come precisely to coincide, it might be supposed that if the fixed one is fired at the exact instant the muzzles come precisely opposite to each other, that the ball from the one will pass down the other. But this from a moment's reflection, is found to be false. The fixed rifle must be fired before the moving one comes opposite, and the allowance must be made by knowing how long the ball requires to move from the one gun to the other, and with what velocity the moving piece is descending. This computation being accurately made, the ball from the shore might be made to enter the muzzle of the moving rifle; but while it is progressing down the barrel, the barrel itself is progressing down the stream, and hence, to avoid the pressure of the ball against the upper side of the barrel, we must fix it in an inclined position, and the bottom of the barrel must be as far up stream as it will descend by the boat's motion during the progress of the ball down the barrel. Hence we see that the direction in which the barrel of the rifle which is to receive the ball is to be placed, is determined by the velocity of the ball, and the velocity of the boat which bear the rifle.

Now for the application. The particles of light coming from the fixed stars are the balls from the fixed rifle. The boat corresponds to the earth sweeping around in its orbit, and bearing with it the tube of the astronomer, down whose axis the particles of light must pass to reach the observer's eye. The velocity of the earth's motion is well known, and the amount by which the telescope must be inclined, to cause the light to enter, has been accurately determined, and from these two data the velocity of the light itself becomes known, and confirms, in the most satisfactory manner, the previously determined value of this incredible velocity, while the reality of the earth's motion is absolutely necessary to render the phenomena at all explicable.

Such were the beautiful results reached by Bradley, and although nothing was gained with reference to the parallax, these preliminary discoveries were in themselves of the highest value, and prepared the way for subsequent observers, who, with better means and more delicate instrumental aid, might prosecute the same great investigation.

Amid the numerous and diversified researches of Sir William Herschel, the problem of the parallax of the fixed stars could not fail to engage his attention by its difficulty and importance. He devised a new means of prosecuting this research, which seemed to promise the most certain success. In his exploration of the heavens with his powerful telescopes, he discovered the curious fact that many fixed stars which, to the unassisted eye, appear as single objects, under the space-annihilating power of the telescope, are seen to consist of two, sometimes of three or more, individual stars, so close to each other that, to the naked eye, they blend into a single object.

Herschel, in the outset, conceived that this proximity of the stars was an accidental circumstance, and that where a pair could be found, in which one individual was about double the other in magnitude, it might reasonably be inferred that the smaller of the two was *twice* as deeply sunk in space as the larger. If this hypothesis could be shown to be true, then would these objects present an admirable means of detecting, with the greatest accuracy, any change in their relative positions, occasioned by the orbitual motion of the earth. In case their proximity was optical, or merely occasioned by the fact that the visual ray drawn to the one passed nearly through the other, it is manifest that, shifting greatly the position of the observer, the stars might be made to open out or close up on each other, or even revolve the one about the other. In employing this mode of investigation, the objects of comparison fell within the field of view of the same telescope, and almost all extraneous sources of error were eliminated.

Such was the plan devised, or rather perfected by Herschel, (for his predecessors had already suggested it), and on the prosecution of which he entered with the zeal which ever distinguished this great astronomer. When he commenced his researches, some half dozen double stars had been discovered and recorded. His first duty was to increase this number as rapidly as possible, and from his entire catalogue to select those best adapted to his purpose. Under his penetrating glance, the number of these curious duplex objects increased with astonishing rapidity, and he was himself startled with their frequent occurrence. However, with a mind fixed on his original design, he selected a large number of pairs, of such relative magnitudes, and in such positions, as promised the most certain success. Let it be remembered that many of these delicate objects were not divided from each other by a space greater than the thousandth part of the sun's diameter.

To ascertain the apparent changes in the relative positions and distances of the stars composing these pairs, Herschel measured, with every care, the distance which separated them, and took the direction of the line drawn from the centre of the one to the centre of the other. Variations of distance and position, occasioned by parallax, were easily computed in kind and character, and the great astronomer commenced and prosecuted his observations with sanguine hopes of success. One thing was certain: —all parallactic movements would have a period of one year, since they arise from the annual revolution of the earth in its orbit, and at the end of this period the stars composing the double sets ought to return to the position occupied at the outset. What was Herschel's astonishment to find that, in many instances the stars composing these pairs were actually in motion; but the movement was certainly not of a parallactic kind, for it neither agreed in direction or in period with the effects of parallax. Here was another grand discovery! These double stars, which were scattered throughout the heavens with far greater profusion than accidental optical proximity could warrant, were found to be magnificent systems of revolving suns! They were united by the law of gravitation, and exhibited the wonderful spectacle of stupendous globes, moving in obedience to the same influences which hold the planets in their orbits, and guide the comets in their eccentric career.—This is not the place to enter into detail concerning these wonderful objects.

While a new field of investigation, boundless and magnificent, was opened up to the human mind; while the great discoverer of these far-sweeping suns was more than rewarded for his toil and labor, the original object of his research was not only left unattained, but the method selected with so much reasonable hope of success, became utterly inapplicable. The parallactic and absolute motions of the systems of stars became so inextricably involved, that the imperfect micrometrical means of Herschel could not separate them.

Thus far, the efforts to obtain the distance of the stars had been un-

THE SCALE UPON WHICH THE UNIVERSE IS BUILT. 153

availing.—A negative solution had indeed been reached. That their distance was enormous, was made evident, from the fact that the parallax had remained insensible, even under the most careful and delicate instrumental tests. Any absolute solution began almost to be despaired of, when hope was again revived by the magnificent refracting telescopes, for which the world was indebted to the skill and genius of the celebrated Frauenhofer, of Munich.—This great artist, aided by the profound science of Bessel, contrived and executed an instrument of extraordinary power, and especially adapted to the research for the parallax of the fixed stars.

Armed with a micrometrical apparatus of wonderful perfection, and capable of executing measures of great, as well as minute distances, the telescope was so arranged as to be carried forward by delicate machinery, with a velocity exactly equal to the diurnal motion of the object under examination.—To give some idea of the delicacy of the contrivances with which these great telescopes have been provided, it is only necessary to state that the micrometer of the great Refractor of the Cincinnati observatory is capable of dividing an inch into 80,000 equal parts !—When mechanical ingenuity failed to construct lines of mathematical minuteness, the spider lent his aid, and it is with his delicate web that these measures are accomplished. Two parallel spider's webs are adjusted in the focus of the eye-piece of the micrometer, and when the light of a small lamp is thrown upon them, the eye, on looking through the telescope sees two minute golden wires, straight and beautiful, drawn across the centre of the field of view, and pictured on the heavens. These are within the control of the observer. He can increase or decrease their distance at pleasure, and so revolve them as to bring them into any position, every motion being accurately measured by properly divided scales.

Suppose, then, it is desired to take the distance and position of the stars forming a pair. The telescope is directed to them, and they are brought to the centre of the field of view. The clock-work is set in action ; it takes up the ponderous instrument, weighing more than 2,500 pounds, and with the most astonishing accuracy it bears it onward, keeping its mighty eye fixed on the object under examination. The observer is thus left with both hands free to make his measures.—He first revolves his micrometer spider's lines round until one of them shall exactly pass from centre to centre of the two stars. This position is noted, and from it is deduced the angle formed by this line with the meridian. He then revolves them a quarter of the circumference, and they are then perpendicular to their former position. He now separates the wires until the one shall exactly bisect one star, while the other wire passes through the centre of the second star, reading this distance on the proper scale. He has fixed, in these two observations, the position and distance of the two components of the double set.—Such is the precision attained in this work, that the most minute motions cannot escape detection. If the stars separate from each other at so slow a rate that a million of years would be required to perform the circuit of the heavens, their motion would be detected in half a year!

With machinery more delicate even than this, and better adapted to the purpose, and of a kind somewhat different, Bessel once more renewed the research after the unattainable parallax of the fixed stars. His great instrument, called the *heliometer*, was mounted as early as 1829, but a multitude of causes, and some unsuccessful efforts, delayed his principal operations up to August, 1837. Three great principles guided him in his selection of 61 in the Swan, as the star on which to perform his observations.—First. It was affected by a very great *proper motion*, a characteristic which we will explain fully hereafter, and which indicated it to be among the nearest of all the stars. Second. Its duplex character adapted it especially to the instrument he was about to employ. Third. The region occupied by 61 Cygni contains a number of minute stellar points, close to the double star, and presenting admirable fixed points, to which the relative motions of the two components of the star to be measured might be referred

With these advantages, and a magnificent instrument, Bessel commenced his observations. He measured the distance from the centre of the line joining the two stars, to two of the small stellar points, which served him as points of reference, and this kind of observation was repeated night after night, whenever the stars were visible, from the middle of August, 1837, up to the end of September, 1838. The entire series of observations was then taken and corrected for every possible known error, and in case any appreciable change remained, it could only be attributed to parallax.

After a most careful and elaborate investigation, a variation commenced to show itself, increasing precisely as parallactic variation ought to increase, and diminishing as it ought to diminish. The period of these changes was precisely a year, and in all particulars, there was an exact correspondence in kind with the changes which ought to be produced by parallax. But such was their minute character, that Bessell hesitated.

During another year the observations were repeated. The same results came out, and the previous values were confirmed. A third year's observations, yielding precisely the same values, removed all doubt, and the great Koeningsburgh philosopher announced to the world that he had passed the impassable gulf of space, and had measured the distance to the sphere of the fixed stars! But how shall I convey any adequate idea of this stupendous distance? Millions and millions of miles serve only to confound the mind. Let us employ a different kind of unit.

Light, as we have seen, travels with a velocity of 12,000,000 of miles in every minute of time. Hence, to reach us from the most remote of all the planets, Neptune, whose distance from the sun is about 3,000,000,000

THE SCALE UPON WHICH THE UNIVERSE IS BUILT. 155

of miles, will require a journey of about four hours; but to wing its flight across the interval which separates our sun from 61 Cygni, will require a period not to be reckoned by hours, or by days, or by months. Nearly ten years of time must roll away before its light, flying, in every second, 192,000 miles, can complete its mighty journey! If the mind revolts at this conclusion; if the distance be too great for comprehension; if the scale of the universe thus suggested even staggers the imagination, I can only say, that all subsequent observation has confirmed, in the most satisfactory manner, the accuracy of Bessel's results. This great astronomer first led the way across the mighty gulf which separates us from the fixed stars. The distance once passed, the route has become comparatively easy, and succeeding observers have determined the parallax of a sufficient number of stars to show that their results are entirely trustworthy.

Having now succeeded in gaining a knowledge of the distance which separates our sun from its remote companions, we are prepared to extend our explorations of the universe. The question naturally arises, how are the stars distributed throughout space?—Are they indifferently scattered in all directions, or are they grouped together into magnificent systems? A cursory examination of the starry heavens with the naked eye shows us, that so far as the large stars are concerned, they do not appear to have been distributed in the celestial sphere according to any determinate law; but on applying the telescope, that luminous zone which, under the name of the Milky Way, girdles the whole heavens, is found to be composed of minute stars, scattered like millions of diamond points on the deep blue ground of the sky.

Sir William Herschel conceived the idea that it might be possible to fathom this mighty ocean of stars, and to determine its metes and bounds; to give to it figure, and to circumscribe its limits. It will not be difficult to explain, in a few words, the general outline of the plan adopted by this extraordinary man in the prosecution of this wonderful undertaking.—In case we admit that the stars are of equal magnitudes, and at equal distances from each other, it would not be difficult to ascertain how far they extended in any given direction, the one behind the other. It is manifest, that in examining the heavens with a telescope of given power and aperture, we shall be able to count more stars in the field of view in those regions where they are so arranged as to reach farthest back into space; and in case we know their absolute distance from each other, the number counted in any field of view, will determine with certainty the length of the visual ray reaching to the most remote star visible in that field.

Now, although the hypothesis that the stars are of equal magnitude, and are uniformly distributed through space, may not be rigorously true, yet doubtless the mean distances are not far from this hypothesis; and although our results may only be approximate, yet as such they are to be relied upon, and they become the more interesting as they carry us to the utmost limits of human investigation. Armed with his mighty telescopes, Sir William Herschel commenced the stupendous task of sounding the heavens, with the purpose of ascertaining whether the stars composing the Milky Way were unfathomable, or were bound and circumscribed by definite limits.

Sweeping a circle round the heavens which cut this grand stratum of stars in a direction nearly perpendicular to its circumference, he directed his great telescope to a certain number of points along this circle, and as he moved slowly onward, counted all the stars visible in each field of view. It was fair to conclude, that wherever most stars were to be seen, there was the stratum deepest. Having gone entirely around the heavens, along the circumference of his circle, he had sounded the depth of the stars along a section of the Milky Way, and to obtain the figure of the section thus cut out was not a difficult matter.

He assumed a central point on paper to represent his point of observation. He then drew from this point lines radiating, and in the actual directions which he had given to his telescope while engaged in his explorations. On each of these indefinite lines he laid off a distance proportioned to the number of stars counted in the field of view in the direction which the line represented, and by joining these points thus determined, he formed a figure which represented the relative depths to which he had penetrated into space; and in case he could be certain that he had gone absolutely through the stratum in every instance, and had grasped every star, even where the extent was most profound, the figure thus constructed would represent the form of the line cut from the outside boundary of the Milky Way by the plane of the circle in which the explorations had been made.

Did he then actually penetrate the deepest portions, or any portion, of the Milky Way? This was now his grand question, and to its decision he gave all his power and ingenuity. As a unit wherewith to measure the space-penetrating power of his telescopes, he assumed the power of the human eye, and knowing that stars of the sixth magnitude are within the reach of the unaided eye, he concluded, from the law regulating the decrease of light, that these minute stars were twelve times more distant than the nearest or brightest stars. Now a telescope having an aperture such as to concentrate twice as much light as the eye, would penetrate into space twice as far, or would reach stars of the twenty-fourth order of distances, and so on for telescopes of all sizes.—In this way he concluded that his great forty foot reflector, with a diameter of four feet, would penetrate 194 times as far as the naked eye, or that it would still see a star of the first magnitude if it were carried backward into space 2,328 times its present distance!

Such, then, was the computed length of the *sounding line* employed in gauging these mighty depths.—Suppose, then, it was required to determine whether this line actually penetrated any given region of the Milky Way. Even with a single telescope, a series of experiments may be performed which go very far to determine this great question. As the space-penetrating power of a telescope depends on the diameter of its aperture, it is easy to give to the same instrument different powers, by covering up, by circular coverings, certain portions of its object glass. Take circles of paste-board, or any other suitable material, and in the first cut an opening one inch in diameter, in the second an opening of two inches, and so on, up to the diameter of the object glass. These diaphrams being successively applied to the object glass, give to the telescope space-penetrating powers proportioned to the diameter of the openings.

In this way Herschel prepared himself to explore one of the deepest portions of the Milky Way. The spot selected was a nebulous or hazy cloud in the sword handle of Perseus, in which, to the naked eye, not a solitary star was visible. I have many times examined the same object, which is certainly one of the most magnificent the eye ever beheld. With the lowest telescopic aid, many stars are rendered visible, surrounded by a hazy light, in which minute glimpse points are occasionally to be seen. As the space-penetrating power was increased, the bright spots of light were successively resolved into groups of brilliant stars, and more nebulous haze came up from the deep distance, indicating that the visual ray was not long enough to fathom the mighty distance. At last the full power of his grand instrument was brought to bear, when a countless multitude of magnificent orbs burst on the sight, like so many sparkling diamonds on the deep blue of the heavens. There was no haze behind; the telescopic ray had shot entirely through the mighty distance, and the clear deep heavens formed the back-ground of the brilliant picture.

Thus did Herschel penetrate to the limits of the Milky Way, and send his almost illimitable sounding line far beyond, into the vast abyss of space, boundless and unfathomable. And now do you inquire the depth of this stupendous stratum of stars? The answer may be given, since we have the unit of measure in the distance of stars of the first magnitude. Light, with its amazing velocity, requires ten years to come to us from the nearest fixed stars, and yet Sir William Herschel concluded, from the examinations he had been able to make, that in some places the depth of the Milky Way was such, that no less than 500 stars were ranged one behind the other in a line, each separated from the other by a distance equal to that which divides our sun from the nearest fixed star.—So that, for light to sweep across the diameter of this vast congeries of stars, would require a period of a thousand years at the rate of 12,000,000 of miles in every minute of time!

The countless millions of stars composing the Milky Way appear to be arranged in the form of a flat zone or ring, or rather stratum, of irregular shape, which I shall explain more fully hereafter. Its extent is so great as properly to form a universe of itself.—If it were possible, to-night, to wing our flight to any one of the bright stars which blaze around us, sweeping away from our own system, until planet after planet fades in the distance, and finally the sun itself shrinks into a mere star, alighting on a strange world that circles round a new and magnificent sun, which has grown and expanded in our sight, until it blazes with a magnificence equal to that of our own, here let us pause and look out upon the starry heavens which now surround us.

We have passed over sixty millions of millions of miles. We have reached a new system of worlds revolving about another sun, and from this remote point we have a right to expect a new heavens, as well as a new earth on which we stand. But no.—Lift up your eyes, and lo! the old familiar constellations are all there. Yonder blazes Orion, with its rich and gorgeous belt; there comes Arcturus, and yonder the Northern Bear circles his ceaseless journey round the pole. All is unchanged, and the mighty distance over which we have passed is but the thousandth part of the entire diameter of this grand cluster of suns and systems; and although we have swept from our sun to the nearest fixed star, and have traveled a distance which light itself cannot traverse in less than ten years, yet the change wrought by this mighty journey, in the appearance of the heavens, is no greater than would be produced in the relative positions of the persons composing this audience to a person near its centre, who should change his seat with his immediate neighbor!

Such, then, is the scale on which the starry heavens are built. If, in examining the magnificent orbits of the remoter planets, and in tracing the interminable career of some of the far-sweeping comets, we feared there might not be room for the accomplishment of their vast orbits, our fears are now at an end. There is no jostling here; there is no interference, no perturbation of the planets of one system by the suns of another. Each is isolated and independent, filling the region of space assigned, and within its own limits, holding on its appointed movements.

Thus far we have spoken only of the Milky Way. In case it be possible to pierce its boundaries, and pass through into the regions of space which lie beyond, the inquiry arises, what meets the vision there? What lies beyond these mighty limits? Does creation cease with this one great cluster, and is all blank beyond its boundary?

Here again the telescope has given us an answer. When we shall have traveled outward from our own sun, and passed in a straight line from star to star, until we shall have left behind us in grand perspective a series of five hundred suns, we then stand on the confines of our own great cluster of stars. All behind blazes with the light of countless orbs, scattered in wild magnificence, while all before us is deep, impenetrable, unbroken darkness. No glance of human vision can pierce the dark profound.

But summoning the telescope to our aid, let us pursue our mighty journey through space; far in the distance we are just able to discover a faint haze of light—a minute luminous cloud which comes up to meet us—and towards this object we will urge our flight. We leave the shining millions of our own great cluster far behind. Its stars are shrinking and fading; its dimensions are contracting. It once filled the whole heavens, and now its myriads of blazing orbs could almost be grasped with a single hand. But now look forward.—A new universe, of astonishing grandeur, bursts on the sight. The cloud of light has swelled and expanded, and its millions of suns now fill the whole heavens. We have reached the clustering of ten millions of stars. Look to the right; there is no limit;—look to the left; there is no end. Above, below, sun rises upon sun, and system on system, in endless and immeasurable perspective. Here is a new universe, as magnificent, as glorious as our own,—a new Milky Way, whose vast diameter the flashing light would not cross in a thousand years. Nor is this a solitary object. Go out on a clear cold winter night, and reckon the stars which strew the heavens, and count their number, and for every single orb thus visible to the naked eye the telescope reveals a *universe*, far sunk in the depths of space, and scattered with vast profusion over the entire surface of the heavens.

Some of these blaze with countless stars, while others occupying the confines of visible space, but dimly stain the blue of the sky, just perceptible with the most powerful means that man can summon to the aid of his vision. These objects are called clusters and nebulæ,—clusters when near enough to permit their individual stars to be shown by the telescope, nebulæ when the mingled light of all their suns and systems can only be seen as a hazy cloud.

Thus have we risen in the orders of creation. We commenced with a planet and its satellite;—we rose to the sun and its revolving planets, a magnificent system of orbs, all united into one great family, and governed by the same great law; and we now find millions of these suns clustered and associated together in the formation of distinct universes, whose number, already revealed to the eye of man, is not to be counted by scores or hundreds, but has risen to thousands, while every increase of telescopic power is adding by hundreds to their catalogue.

Let us now explain these "island universes," as the Germans have aptly termed them, and attempt approximately to circumscribe their limits, and measure their distances from us, and from each other.—Sir William Herschel, to whom we are indebted for this department of astronomy, conceived a plan by which it was possible approximately to sound the depths of space, and determine, within certain limits, the distance and magnitudes of the clusters and nebulæ within the reach of his telescope. To convey some idea of his method of conducting these most wonderful researches, imagine a level plane, of indefinite extent, and along a straight line, separated by intervals of one mile each, let posts be placed, bearing boards on which certain words are printed in letters of the same size. The words printed on the nearest board, we will suppose, can just be read with the naked eye. To read those on the second, telescopic aid is required, and that power which suffices to enable the letters to be distinctly seen, is exactly double that of the unaided eye. The telescope revealing the letters at the distance of three miles is three-fold more powerful than the eye, and so of all the others. In this way we can provide ourselves with instruments whose space-penetrating power, compared with that of the eye, can be readily obtained.

Now to apply these principles to the sounding of the heavens. The eye, without assistance, would follow and still perceive the bright star Sirius, if removed back to twelve times its present distance.—After this, as it recedes, it must be followed by the telescope. Suppose, then, a nebula is discovered with a telescope of low power, and it is required to determine its character and distance. The astronomer applies one power after another, until he finally employs a telescope of sufficient reach to reveal the separate stars of which the object is composed, which shows it to be a cluster; and since the space-penetrating power of this instrument is known, relative to that of the human eye, in case the power is one hundred times greater than that of the eye, then would the cluster be located in space one hundred times farther than the eye can reach, or twelve hundred times more remote than Sirius, or at such a distance that its light would only reach our earth after a journey of 120,000 years!

Such was Herschel's method of locating these objects in space. Some are so remote as to be far beyond the reach of the most powerful instruments, and no telescopic aid can show them other than nebulous clouds of greater or less extent. It was while pursuing these grand investigations that Herschel was led to the conclusion, that among the nebulæ which were visible in the heavens, there were some composed of chaotic matter, a hazy, luminous fluid, like that occasionally thrown out from comets on their approach to the sun.

Among these chaotic masses he discovered some in which the evidences of condensation appeared manifest, while in others he found a circular disk of light, with a bright nucleus in the centre. Proceeding yet farther, he found well formed stars surrounded by a misty halo, which presented all the characteristics of what he now conceived to be nebulous fluid. Some of the unformed nebulæ were of enormous extent and among those partially condensed, such as the nebulæ with planetary disks, many were found so vast that their magnitude would fill the space occupied by the sun and all its planets, forming a sphere with a diameter of more than 6000 millions of miles. Uniting these and many other facts, the great astronomer was finally brought to believe, that worlds and systems of worlds might yet be in the process of formation, by the gradual conden-

THE SCALE UPON WHICH THE UNIVERSE IS BUILT.

sation of this nebulous fluid, and that from this chaotic matter originally came the sun and all the fixed stars which crowd the heavens. This theory, extended, but not modified, in the hands of Laplace, is made to account for nearly all the phenomena of the solar system, and has been already referred to in a former lecture.

For a long time, this bold and sublime speculation was looked upon, even by the wisest philosophers, with remarkable favor. The resolution of one or two nebulæ, (so classed by Herschel), with the fifty-two feet reflector of Lord Rosse, has induced some persons to abandon the theory, and to attempt to prove its utter impossibility. All that I have to say, is, that Herschel only adopted the theory after he had resolved many hundreds of nubulæ into stars; and if there ever existed a reason for accepting the truth of this remarkable speculation, that reason has been scarcely in any degree affected by recent discoveries.

I have examined a large number of these mysterious objects, floating on the deep ocean of space like the faintest filmy clouds of light. No power, however great, of the telescope, can accomplish the slightest change in their appearance. So distant that their light employs (in case they be clusters) hundreds of thousands of years in reaching the eye that gazes upon them, and so extensive, even when viewed from such a distance, as to fill the entire field of view of the telescope many times. Sirius, the brightest, and probably the largest of all the fixed stars, with a diameter of more than a million of miles, and a distance of only a single unit, compared with the tens of thousands which divide us from some of the nebulæ; and yet this vast globe, at this comparatively short distance, is an inappreciable point in the field of the telescope. What, then, must be the dimensions of those objects, which, at so vast a distance, fill the entire field of view even many times repeated?

Herschel computes that the power of his great reflector would follow one of the large clusters if it were plunged so deep in space that its light would require 350,000 years to reach us, and the great telescope of Lord Rosse would pursue the same object probably to ten times this enormous distance.

Such examinations absolutely overwhelm the mind, and the wild dream of the German poet becomes a sort of dreadful sublime reality:—

" God called up from dreams a man into the vestibule of heaven, saying, ' Come thou hither, and see the glory of my house.' And to the servants that stood around his throne he said, ' Take him, and undress him from his robes of flesh : cleanse his vision, and put a new breath into his nostrils; only touch not with any change his human heart—the heart that weeps and trembles.' It was done: and, with a mighty angel for his guide, the man stood ready for his infinite voyage; and from the terraces of heaven, without sound of farewell, at once they wheeled away into endless space. Sometimes with the solemn flight of angel wing they fled through Zaarrahs of darkness, through wildernesses of death, that divided the worlds of life; sometimes they swept over frontiers, that were quickening under prophetic motions from God. Then, from a distance that is counted only in heaven, light dawned for a time through a sleepy film; by unutterable pace the light swept to *them*, they by unutterable pace to the light. In a moment the rushing of planets was upon them: in a moment the blazing of suns was around them.

"Then came eternities of twilight, that revealed, but were not revealed. On the right hand and on the left towered mighty constellations, that by self-repetitions and answers from afar, that by counter-positions, built up triumphal gates, whose architraves, whose archways—horizontal, upright—rested, rose—at altitude by spans—that seemed ghostly from infinitude. Without measure were the architraves, past numbers were the archways, beyond memory the gates. Within were stairs that scaled the eternities below; above was below—below was above, to the man stripped of gravitating body: depth was swallowed up in height insurmountable, height was swallowed up in depth unfathomable. Suddenly as thus they rode from infinite to infinite, suddenly, as thus they tilted over abysmal worlds, a mighty cry arose—that systems more mysterious, that worlds more billowy,—other heights and other depths,—were coming, were nearing, were at hand.

"Then the man sighed, and stopped, shuddered, and wept. His overladened heart uttered itself in tears; and he said—'Angel, I will go no farther. For the spirit of man acheth with this infinity. Insufferable is the glory of God. Let me lie down in the grave and hide me from the persecution of the infinite; for end, I see there is none.' And from all the listening stars, that shown around issued a choral voice, 'The man speaks truly: end there is none, that ever yet we heard of.' 'End is there none?' the angel solemnly demanded: 'Is there indeed no end?—and is this the sorrow that kills you?' But no voice answered, that he might answer himself. Then the angel threw up his glorious hands to the heaven of heavens, saying, 'End is there none to the universe of God. Lo! also there is no beginning.'"

LECTURE X.

THE MOTIONS AND REVOLUTIONS OF THE FIXED STARS.

HAVING reached, in the course of the preceding lecture, to the outermost confines of the visible creation, let us now return home from this survey of the "island universes" which crowd the illimitable regions of space, to the stars which compose our own cluster, and learn how far the human mind has progressed in its examination of the millions of suns which constitute, in a more definite sense, our own Milky Way.

We have already seen that the parallax of 61 Cygni rewarded the laborious and extraordinary efforts of Bessel. The example set by this great astronomer encouraged those who followed him, and while his results in this particular case had been confirmed in the most astonishing manner, the distances of many other stars have been obtained, until a sufficient amount of data has been accumulated to determine the approximate distances of the sphere of the fixed stars of different magnitudes. Struve estimates the mean distance of stars of the first magnitude to be 986,000 times the radius of the earth's orbit, or so remote that their light reaches us only after a journey of fifteen years and a half. Stars of the second magnitude send us their light in twenty-eight years, those of the third magnitude in forty-three years; while the light from stars of the ninth magnitude only reaches the eye of the observer after traversing space for five hundred and eighty-six years, at the rate of twelve millions of miles in every minute of time.

My range of investigation does not permit me to explain, at this time, how these extraordinary conclusions have been reached, The reasoning, however, is close and clear, and the results are no doubt approximately correct.

Such, then, are the distances separating man from the objects of his research. To have attained to a knowledge of these distances even, is sufficiently wonderful, but what we are about to reveal as the results of human investigation among these far distant orbs, cannot fail to fill the mind with astonishment, and demonstrate the great truth that "man has been made but a little lower than the angels."

Before it became possible to examine with absolute certainty the places of the stars, with a view to ascertain their absolute fixity, many difficult preliminary preparations had to be accomplished. Instruments of the most perfect kind must be provided, not only in their optical performances, but in their space-dividing machinery. Moreover, the places of the stars, as determined by the best telescopes, must be corrected for every possible instrumental error. The two points to which the stars are referred are the north pole and the vernal equinox. In case any motions belong to these points, their amounts and directions must be ascertained and allowed for. Then the effects of refraction, and of the abberration of light, were indispensable to a perfect investigation of the absolute places of the stars.

All these and many other preliminary matters having been satisfactorily determined, it became possible to examine, in the most critical manner, the places of the stars, and to learn whether indeed, (as had been supposed for thousands of years), their configurations were eternal and unchangeable, or whether they moved among themselves with a motion rendered so slow by their immense distance, as hitherto to have escaped the most scrutinizing watch.

Fully armed with the necessary instruments, it did not require many years to determine the grand truth, that among the tens of thousands of stars which fill the heavens, not a solitary one, in all probability, is in a state of absolute rest. Many were found to move so swiftly, that their velocity was determined even in a single year; while others, in consequence of their enormous distance, may require centuries to detect any appreciable change. In the outset these extraordinary movements seemed to be directed by no law—some stars were sweeping in one direction, and some in another. Motion, ceaseless, eternal motion, seems to be stamped on the entire universe, and while the stars are pursuing their mighty orbits, we cannot resist the idea that our own sun, the centre of our great planetary system, itself a star, must participate in the general movement, and is, in all probability, urging its flight, accompanied by all its planets, satellites, and comets, to some unknown region of space.

The revolution of the stars, the organization of the grand cluster with which our sun is associated, the demonstration of the sun's absolute translation through space, its direction, velocity, and period, are the topics to which I invite your attention in the closing lecture of the present course.

When forced to acknowledge the rotation of our globe on its axis, and its swift orbitual motion, surrounded by wheeling planets and flying comets, the mind naturally retreats to the sun as the great immovable centre, where it can rest and contemplate these circling worlds. But even here, as we shall presently see, there is no rest. The sun himself becomes a subordinate member of a grander combination of worlds, and, obedient to higher influence, sweeps around in its unmeasured orbit.

We shall present a rapid summary of the evidence of change among the fixed stars, and then proceed to develope the reasoning by which the direction and velocity of the sun's motion in space has been determined.

More than two thousand years ago, the celebrated Greek astronomer,

Hipparchus, was astonished by the sudden bursting forth of a brilliant star in a region of the heavens where none had previously existed. Up to this time, no doubt of the immutability of the starry sphere seems to have been entertained, and while the philosopher gazed and wondered, he resolved to execute a work from which posterity might learn the change of the celestial sphere. He undertook and completed his great catalogue of the places of a thousand stars, locating them with all the accuracy permitted by the rude instruments then in use. Subsequent observers, by comparing their own determined positions of the stars with their places as fixed on the catalogue of Hipparchus, could readily perceive any sensible change which might occur in their configuration, the appearance of new stars, or the disappearance of those which had once existed.

The sudden breaking forth of a new star is a phenomenon of such wonderful character that we might well doubt the possibility of its occurrence, if we were obliged to rely on the historical account transmitted to us from the time of Hipparchus. But, fortunately, more than one brilliant example of the kind has occurred in modern times, presenting the most unequivocal evidence of the reality of this inexplicable wonder.

In 1572, a new star of great splendor appeared suddenly in the constellation Cassiopeia, occupying a position which had previously been blank. This star was first perceived by Schuler, of Wittemburg, on the 6th of August. It was detected by Tycho, the Danish astronomer, on the 11th of the following November, and the wonder produced by this most extraordinary phenomenon induced him to give to the star the most unremitting attention. Its magnitude increased until it is said to have surpassed even Jupiter in splendor, and finally became visible in the day-time. It retained its greatest magnitude but for a very short time, when it commenced to diminish in brilliancy, changing from white to yellow, then to reddish, and finally it became faintly blue; and so diminishing by degrees, it vanished from the sight in March, 1574, and has never since been seen.

In the year 1604, while the scholars of Kepler were engaged in observations of Mars, Jupiter and Saturn, then in close proximity to each other, having been interrupted a day or two by clouds, on the return of fine weather, Maestlin was astonished to find near the planets then in the constellation Ophiuchus, a brilliant star, which certainly had not been there a few days before. This object attracted the attention of all the great astronomers then living, and was particularly observed by Galileo and Kepler. It is said to have attained a splendor equal to that of the planet Venus, and from this, its greatest brilliancy, it gradually declined, until, about the beginning of 1606, it ceased to be visible, and no telescopic power has since been able to detect any star in the place once occupied by this remarkable stranger.

Although observed with the greatest care, no sensible parallax was ever detected in either of these objects, and no doubt exists as to their occupying the region of the fixed stars. Many other less remarkable examples are on record, but up to the present no satisfactory explanation of this astonishing phenomenon has been given, Whether it indicates the actual destruction of some magnificent system, or the revolution of these stars in orbits of great eccentricity, causing them to appear to us, like the comets, only in the perihelion points of their mighty orbits, is equally uncertain. One thing is certain: they present evidence of change in the starry heavens, of the most startling and irresistible kind.

While new stars have occasionally made their appearance, to astonish mankind with their brilliancy, there are many well authenticated cases of the entire disappearance of old stars, whose places had been fixed with a degree of certainty not to be doubted.—In October, 1781, Sir William Hershel observed a star, No. 55 in Flamsted's catalogue, in the constellation Hercules. In 1790, the same star was observed by the same astronomer, but since that time no search has been able to detect it. The star is gone; whether never to return, it is impossible to say. A like disappearance has occurred with reference to the stars numbered 80 and 81, both of the fourth magnitude, in the same constellation. In May, 1828, Sir John Herscchel missed the star numbered 42, in the constellation Virgo, which has never since been seen. Examples might be multiplied, but it is unnecessary.

In these cases the stars have been lost entirely;—no return has ever been marked; and but for the discovery of another class of remarkable objects among the stars, no return would probably ever have been suspected. If I could direct your attention to-night to a brilliant star named Algol, in the head of Medusa, and bring a powerful telescope to aid in your examinations, this star, while you are watching it, might be seen to lose its splendor, and from its rank of the second magnitude to decline in brightness, until it would scarcely be visible to the naked eye. Having reached a certain limit, it would commence an increase, and by slow degrees resume its original splendor.—This decrease and increase is actually accomplished in about eight hours. Having regained its usual light, it remains stationary for about two days and a half, and then repeats the changes already detailed; and thus have its periodical fluctuations continued since the date of its discovery, with the most astonishing regularity. The bright star marked Beta, in the constellation Lyra, is known to pass from the third to the fifth magnitude, and to regain its light in a period of six days and nine hours. These are called *periodical* stars and a sufficient number have already been detected to present a progressively increasing series of periods from two days twenty hours up to four hundred and ninety-four days, and in one case even many years.

Here, again, are phenomena indicative of extraordinary activity in these remote regions of space.—No explanation of these changes has yet been given in all respects satisfactory. Some have attributed them to the existence of dark spots on the stars, which, by rotation on an axis, periodically present themselves, and thus dim the lustre of the stars. Others think the changes are due to the revolution of large planets about the stars which, by coming between the eye and the star, eclipse a portion of its light; while a third class conceive the fluctuations to arise, in some instances at least, from an orbital motion of the stars in orbits of excessive elongation, and so located as to have their greater axes directed towards our system.

It will be seen that this theory may be readily extended so as to embrace the new stars already referred to, and even to account for those which have been lost from their places in the heavens. Here, however, we enter the confines of the uncertain. Centuries may roll away before the true explanation of these astonishing changes shall be given; but the mind is on the track, and with a steady and resistless movement is slowly pushing its investigations deeper and still deeper into the dark unknown.

While the phenomena of the new and lost stars, and the fluctuations in the light of the variable ones, gave undeniable evidence of constant change in what Aristotle was pleased to call the eternal and incorruptible heavens, Herschel's brilliant discovery of the orbital motion of the double stars gave to the mind the opportunity of determining the nature of the law which sways the movements in these distant regions of space. It was natural, in the first efforts to compute the orbits of the double stars, to adopt the hypothesis that they attracted each other by the same law which prevails in the planetary system. Results did not disappoint expectation.— Gravitation, which Newton, in the outset of his great discovery, had boldly affirmed exerted its influence wherever matter existed or motion reigned, was extended, in the most absolute manner, to the region of the fixed stars. There, at a distance from our own system almost inconceivable, suns and systems of suns, rising in orders of greater complexity, revolving with swift velocity, or with slow and majestic motion, bore testimony, ample and unequivocal, to the truth of the great law of universal gravitation.

Every particle of matter in the universe attracts every other particle of matter with a force which is proportioned directly to the mass, and which decreases as the square of the distance at which it operates increases. This is no longer a bold hypothesis. The double star marked Zeta, in the constellation Hercules, has been subjected to the analysis of the computer. The elements of its orbit have been obtained, and true to its predicted period, it has actually performed an entire revolution in a period of thirty-five years. The components of the star Eta, in the Northern Crown, revolve around their common centre in about forty-four years. Both of these pairs have completed an entire revolution since their discovery. Many others might be named, but my only object, at present, is to exhibit the evidence which shall remove all doubt as to the actual extension of the law of gravitation to the fixed stars.

Let it be remembered that this department of astronomy is yet in its infancy. Thousands of double stars have been detected, and every year adds hundreds to the list. Among these, a large proportion must prove to be binary systems, varying in their periods of revolution, from thirty years or less, up to many thousands, perhaps millions of years.

The association of two suns naturally suggests the possible union of a greater number, forming more complicated systems. This idea has been verified—a large number of triple systems has been discovered. In a few instances quadruple sets have been found, of which a remarkable example exists in the constellation of the Harp. Here was found four suns, arranged in pairs of two. The components of the first pair revolve around each other in about one thousand years; those of the second pair appear to require about double that period, while one pair revolves about the other in a period which, determined roughly from their distance, cannot fall much below a million of years! The evidence of the physical union of these four stars into one grand system rests, at present, on the ascertained fact that their proper motions are the same.

From quadruple systems we rise, by analogy, still higher, until we find hundreds, sometimes thousands, of stars compacted together in so small a compass that their proximity cannot be the effect of accident. Look at the beautiful little cluster called the Pleiades: an ordinary eye may here see six or seven stars. One of very great power has been known to count fourteen in this group, while the telescope increases the number to hundreds; and yet the space in which they are located might easily be covered by the moon.

Suppose an indifferent scattering of the stars through space, and compute the chances that such a number would fall together at any one point, and we shall find not one chance out of millions in favor of such an accident. We are therefore forced to the conclusion that here is a more magnificent order, one in which hundreds of suns, surrounded by their subordinate worlds, are all united by gravitation into one grand system This is not a solitary example.—Many of these beautiful objects, comparatively close to our sun, are found in the heavens, leading the mind gradually up to the contemplation and examination of that mighty system of systems, that great cluster of clusters, the Milky Way, of which all these are but subordinate groupings,—vast in themselves, but when compared with the whole, mere units among the millions of which it is composed.

From what we have seen, it is impossible to avoid the conclusion that gravitation exerts its power among the myriads of shining orbs which strew the Milky Way. The innumerable suns which form this stupendous cluster must feel the reciprocal influence of each other, and nothing short of the centrifugal force arising from orbital motion can balance this universal attractive power, and give to this grand system the great characteristic of stability.

Herchel succeeded, at least approximately, in sounding the profundities of the Milky Way, and fixed the relative position of our own sun among the stars by which it is surrounded. He found it to be located not very distant from the centre of the great stratum and near the line where the principal current of stars divides into two great streams, which for a time separate from each other, but finally reunite in a distant region of the heavens.

Having accomplished thus much, this great astronomer attempted the resolution of the grand problem of the sun's movement through space. This investigation is so lofty, so daring and utterly incomprehensible at the first glance, that its mere announcement produces little effect on the mind. Consider, for one moment, what it involves. Man is located on a planet almost infinitely larger than himself. This planet is swiftly revolving on its axis, and in its orbit round a great central luminary, the sun. The daring philosopher participates in all these motions. He provides himself with instruments which measure the distances and positions of the almost infinitely distant fixed stars. These fixed stars, when subjected to his critical examination, cease to be fixed, and are found to be moving with astonishing velocity in all directions. Among these he numbers his own sun, and although borne along in the progressive motion of his own great centre, he ventures to attempt the determination of the fact of its actual motion, the direction in which it moves, and the velocity with which it is sweeping through space.

This problem is so wonderful that I beg your earnest attention while the effort is made to simplify the reasoning by which its resolution has been accomplished.

Before the actual motions of the earth were discovered, the sun, moon, and planets, as well as the stars, *appeared* to move in certain directions, and with certain velocities, not easily explained. The rotation of the earth on its axis rendered a clear explanation of the diurnal movements of the heavenly bodies, and its orbital motion around the sun explained the sun's apparent annual movement among the fixed stars. Thus it is seen and readily apprehended, that in case the spectator is progressing, his actual motion may be transferred to distant bodies under examination, and these may appear to move while he seems to be at rest.

Now in case the sun is sweeping towards any quarter of the heavens, it must carry with it all its planets, satellites, and comets. The earth is borne along in common with its companions, and the observer on its surface will transfer his own movement through space to the distant objects which only appear to change their places, in consequence of his own translation through space. Thus the distant stars may be affected with a parallactic change, not to be confounded with that produced by the revolution of the earth in its orbit, but occasioned by the fact that while the earth revolves around the sun, she is carried forward by this luminary in his journey through space. As the whole system participates in this motion, in case the planets are inhabited, their astronomers will detect in the fixed stars the parallactic motion due to the sun's movement, and hence this change among the stars may be properly termed their *systematic parallax*.

Herschel commenced his examination of this great problem by forming a catalogue of stars situated in all parts of the heavens, in which an appreciable amount of *proper motion* had been detected and measured. Now in case this apparent motion of the stars could be attributed to the movement of the solar system through space, a close scrutiny of the directions in which the stars appeared to move would indicate the direction in which the observer, carried along with the sun, was passing through space.

In case a person is travelling on a railway, in a direct line through a forest of trees, as he advances, all objects towards which he is moving will appear to open out or separate from each other, while those left behind will appear to close up. If, then, the astronomer, borne along by the movement of the sun through the vast *forest* of stars by which he is surrounded, desires to ascertain the direction in which he is progressing, let him search the heavens until he finds a point where the stars seem to be increasing their distance from each other. Should he find such a point, let him confirm his suspicions by looking in the direction precisely opposite and behind him, and in case he finds the stars located in this region closing up on each other, he may fairly conclude that he has found the direction in which he is moving, and a rigid coincidence of all the phenomena would demonstrate the accuracy of his conclusions.

Such was the general train of investigation adopted by Herschel. After as extended an examination as the data with which he was then furnished permitted, he announced his belief that a part of the proper motion of the fixed stars must be attributed to the effect of systematic parallax, and that the solar system was moving through space towards a point in the constellation Hercules.

The announcement of this astonishing result was received with hesitation and doubt by the best living astronomers, and Herschel died before any confirmation of his great theory had been obtained. After his death, for nearly half a century, no mind seemed willing to renew the investigation. The theory fell into disrepute, and was only regarded as a bold and sublime speculation, but not founded on any well determined observations.

Within a few years, the problem has engaged the attention of the distinguished astronomers of Russia. Argelander, of Bonn, led the way, and by a train of reasoning based upon extensive and accurate observations, has sustained and demonstrated, in the most undeniable manner, not only the general truth of Herschel's theory, but has even confirmed the direction in which that astronomer believed the solar system to be moving.

Here again permit me to attempt a popular explanation of Argelander's reasoning. Suppose a single star to have its place fixed absolutely by observation on the first day of the year 1700. One hundred years after, its place is again determined, when it is found to have shifted its position. Conceive the star to have so moved as to reach the meridian earlier than it formerly did. When on the meridian, its old place will be behind or east of the new place, and a line joining the old and new places will show the direction in which the star has been moving, and the distance between the two places will exhibit the amount of motion in one hundred years. If the star do not move exactly north or south, its line of direction will form an angle with the meridian, whose value is determined from a comparison of the old and new places of the star.

Argelander commenced by selecting five hundred stars, in all regions of the heavens, whose places had been well determined by preceding astronomers. The preference was of course given to those which had been longest subjected to observation. Having himself determined the new places of all these stars, a comparison of his own with previously observed positions determined the direction in which these stars were moving, and their rates of motion. The angles formed by the lines along which each star was progressing, with the meridian, became known from observation, and these angles we shall call the *observed angles of direction*. Now it is not difficult to compute the directions in which the stars would appear to move, if their motion be produced by the movement of the solar system.

Suppose, for example, that the sun, with its planets, is sweeping exactly towards the north pole of the heavens, then would all the stars appear to move towards the south. Those in the equator would move with the swiftest velocity from the north pole, but those nearest the pole would appear to separate from each other, while their recess from the pole would be comparatively slight. To render this reasoning still plainer, imagine this room to be pierced on every side, so that an eye placed at the centre could see every star in the heavens through the openings. Through each of these holes conceive iron rods to pass, all meeting at a point in the centre, and all directed exactly to the stars. On the outside let golden balls be fixed to the extremities of these rods, to represent the stars. Now, grasping the extremities of all these rods in the hand, urge the point where they all unite towards the north pole, and watch the movement of the balls at the outer extremities of the rods. The ball corresponding to the north star will scarcely seem to move, because the eye travels directly towards it. The balls corresponding to the stars on the equator, having their rods perpendicular to the direction of the motion of the central point will sweep swiftly towards the south. The idea once gained, there is no difficulty in its application.

The visual rays drawn to the stars correspond to the rods, and these rays, meeting in the eye of the observer, are carried forward by the sun in its progression through space. I have supposed the system to move due north; but in case the motion be assumed in any other direction, it is easy to compute the changes consequent. Understanding these preliminary statements, we are prepared to follow Argelander in his investigation.

The five hundred stars selected for examination were divided into three groups, according to the amount of annual proper motion. The first contained only such stars as were seen to move with a velocity not less than one second of space in a year. Although this motion may appear excessively slow, yet its direction in one hundred years may be determined with very great precision. A general examination of the direction in which the stars of this first group appeared to move, indicated the quarter of the heavens towards which the solar system must be progressing; and now commenced the investigation, having for its object the discovery of the exact point.—To accomplish this, a point was assumed, and on the hypothesis that it was correctly chosen, the directions of the motion of all the stars composing the first group were computed, and the angles formed by their lines of direction with the meridian were determined.

If the motion of these stars was the effect of systematic parallax, and if the direction of the solar movement had been accurately chosen, then would the *computed angles of direction* agree exactly, in every instance, with the *observed angles of direction*. The comparison of these angles having been made, it was easy to see the discrepancies, and by shifting the assumed point, these differences could be reduced to their minimum value. The point which gave the smallest differences between the observed and computed angles would be the one towards which the solar system was progressing. Such was the reasoning of Argelander, and such the train of investigation on which he relied for the resolution of this great problem.

Having closed his examinations based on the group of stars with the most rapid motion, and having found the point in the heavens which corresponded to their motions, he proceeded to execute his calculations with reference to his second group. The stars of this group moved annually an amount greater than half a second of space, and less than one second. The result was again reached, and direction of the solar motion thus derived, agreed, in a remarkable manner, with that obtained from the first group. A further confirmation was obtained by executing the calculation founded on the motions of the third and last group into which he had divided his five hundred stars.—The final result settled, probably forever, the grand fact that the sun, with its entire cometary and planetary system, is sweeping through space towards a point whose place must fall somewhere within the circumference of a circle whose diameter is about equal to four times that of the moon.

The reality of the solar motion once determined, astronomers have not been wanting to verify and extend this wonderful examination. Argelander's results have been confirmed by the investigations of M. Otho Struve, the son of the distinguished director of the Imperial Observatory of Pulkova; and if, on any fair night, you direct your eye to the constellation Hercules, and select from its stars the two marked on the globe with the Greek letters π and μ on the line joining these stars, and at a distance from π equal to one-quarter of the distance which divides the stars will be found the point towards which the sun was directing his course in the year 1840.

Having obtained the direction of the solar motion, we proceed to investigate its actual volocity. How swiftly does the sun, with its retinue of worlds, sweep onward through space? It will not be possible to present here even an outline of the reasoning of Struve in the resolution of this intricate question.—Two points are involved. The determination of the annual angular motion of the sun, as it would be seen by a spectator situated at a distance equal to that of the stars of the first magnitude. This being determined, the angular motion can readily be converted into linear velocity, in case the mean distance of the stars of the first magnitude can be satisfactorily obtained. After an elaborate investigation, guarded by every care, and open, as it would appear, to no well founded objections, M. Otho Struve has finally resolved the first of these wonderful questions. It is curious to see how nearly the results agree, which were obtained from data entirely different, and in no way dependent on each other.

By an examination based on observed right ascensions of the stars, he finds that the space passed over by the sun in its progressive movement through the heavens, seen from the mean distance of the stars of the first magnitude, is three hundred and twenty-one-thousandths of a second of arc. The result obtained from observed declinations gave for the same quantity three hundred and fifty-seven-thousandths of one second of arc. Here is a difference amounting to only thirty-six-thousandths of a second, a quantity exceedingly small, when we consider the extraordinary difficulty of the investigation.

Let us now convert these members into intelligible quantities. In case the sun be supposed to be revolving about some mighty centre, at a distance equal to the mean distance of stars of the first magnitude, the period necessary to accomplish its stupendous revolution will be 3,811,000 years!

Vast as this period appears, we shall see hereafter that we have no right to suppose that the centre about which the solar system is revolving, can be located at a distance nearly so small as the mean distance of the larger stars. But what is the actual velocity?—How many miles does this mighty assemblage of flying worlds accomplish in its unknown journey in every year? This is the last question, and even this has not escaped the successful examinations of the human mind. The discovery of the parallax of one or two fixed stars has already been referred to.—Within a few months, an elaborate work, by Struve, on the Sidereal Heavens, has reached us, containing some remarkable investigations on the mean distances of the stars of the various magnitudes.

Struve, by a most ingenious and powerful train of investigation, obtains a series representing the *relative* mean distances of the stars of all magnitudes, up to the most minute visible in Herschel's twenty feet reflector. From the sun, as a centre, he sweeps successive concentric spheres, between whose surfaces he conceives the stars of the several magnitudes to be included. The radius of the first sphere reaches to the *nearest* stars of the first magnitude; that of the second sphere extends to the *farthest* stars of the same magnitude, and the means of these two radii will be the mean distance of the stars of the first magnitude. The same is true with reference to the concentric spheres embracing within their surface the stars of the various orders of brightness.

Having, from his data, computed a table exhibiting the relative distances of the stars of the different magnitudes, an examination of these figures revealed the singular fact that they constituted a regular geometrical progression; and having assumed the distance of the stars of the sixth magnitude as the unit, the distance of the stars of the fourth magnitude will be *one-half*; that of those of the second magnitude will be *one-quarter*, and so to the even numbers expressing magnitude; while the distance of the stars of the *fifth* magnitude is obtained by dividing unity by the square root of the number 2, and from this the distance of the odd magnitudes come by dividing constantly by 2. In mathematical language, the distances of the stars of the various magnitudes form a geometrical progression whose ratio is equal to unity divided by the square root of 2.

Having thus obtained the *relative* mean distances of the stars, in case we can find the absolute mean distances of those of any one class, that will reveal to us the absolute mean distances of the stars of every class. For the approximate accomplishment of this last great object, we are again indebted to the astronomers of Russia. As early as 1808, M. Struve, then of Dorpat, attempted the determination of the parallax of a large number of stars, and obtained results so small that, in the state of astronomical science as it then existed, no confidence could be placed in them. The final value of the numerical co-efficient of the aberration of light had not been then absolutely determined. Subsequent investigations by Struve and Peters have fixed this quantity, and the actual determination of the parallax of eight stars recently, has shown that confidence may now be placed in the results obtained by Struve nearly 25 years ago.

By combining all the results, M. Peters finds no less than thirty-five stars whose parallaxes have now been determined, either absolute or relative, with a degree of accuracy which warrants their employment in investigating the problem of the mean parallax of stars of the second magnitude. Excluding from this number the stars 61 Cygni, and No. 1830 of the Grombridge catalogue, on account of their great proper motion, there remained thirty-three stars to be employed in the investigation.

From a full and intricate examination of all the data, by a process of reasoning which I will not attempt to explain at this time, M. Peters finds the mean parallax of stars of the *second* magnitude to be equal to 116 thousandths of one second of arc, with a probable error less than a *tenth* part of this quantity. Returning now, with this absolute result, to the table of the relative distances of the fixed stars of different magnitudes, it is easy to fix their absolute distances, as far as confidence can be placed in this first approximation. We find the stars of the first magnitude to be located between the surface of two spheres, whose radii are respectively nine hundred and eighty-six thousand times the radius of the earth's orbit, and one million two hundred and forty-six thousand times the same unit. We will express the distance in terms of the velocity of light, as no numbers can convey any intelligible idea. Stars of the first magnitude send us their light in about seventeen years; those of the second magnitude in about thirty years;—stars of the third magnitude send their light in about forty-five years; those of the fourth magnitude in sixty-five years; those of the fifth in ninety years; those of the sixth magnitude, the most remote visible to the naked eye, send us their light after a journey through space of one hundred and thirty years! while the distance of the lowest order of telescopic stars visible in Herschel's twenty feet reflector is such, that their light does not reach the eye for 3,541 years after it starts on its tremendous journey!

Let it be remembered that these results are not conjectures. Though they are first approximations to the truth, they are reliable to within the tenth part of their value, and are thus far certain; they raise, in the most astonishing manner, our views of the immensity of the universe, and of the powers of human genius which have fathomed these vast and overwhelming profundities.

Let us now return to the examination of the absolute amount of progressive motion of our sun and system through space. As already stated, M. Otho Struve determined its yearly angular motion, as seen from the more distant of the stars of the first magnitude. To convert this angular motion into miles, a knowledge must be obtained of the absolute mean distance of the stars of the first magnitude. This has been accomplished by M. Peters, and combining the researches of Argelander, Struve, and Peters, we are now able to pronounce the following wonderful results.— *The sun, attended by all its planets, satellites, and comets, is sweeping through space towards the star marked π in the constellation Hercules, with a velocity which causes it to pass over a distance equal to thirty-three millions three hundred and fifty thousand miles in every year!*

And now do you demand how much reliance is to be placed on this bewildering announcement? I answer, that as to the reality of the solar motion, there is but one chance out of four hundred thousand that astronomers have been deceived. We cannot resist the evidence, and startling as the truth appears, we are obliged to yield our assent, reluctant though it may be, to the logical reasoning by which this magnificent result has been demonstrated.

But whither is our system tending? If moving onward with such tremendous velocity, is there not danger that ere long it may reach the region of the fixed stars, and by sweeping near to other suns and systems, derange the order of the planetary worlds? Let us examine this question for one moment, on the hypothesis that the sun alone is moving among all the stars of heaven, and that it will hold on in its present direction until it shall reach the star in Hercules, towards which it is now urging its flight. This star is of the third magnitude, and according to our statement already made, the mean distance of its class is such, that its light does not reach us in a period less than forty-six years. Executing the calculation, we find that in case the solar system should continue to progress towards that star, it cannot pass the enormous interval, even at 33,550,000 miles per annum, in less than 1,800,000 years!

If the eye of any superior intelligence can behold this amazing scene, how stupendous must be the spectacle presented! In the centre the sun, blazing with splendor, pursues its majestic career;—around it roll the planets, and about it cluster ten thousand fiery comets. Worlds bright and beautiful hover near the sun,—worlds fiery and chaotic seek this great centre with impetuous velocity, and then dash away into the farthest range of their grand revolution. But the monarch moves on, and his magnificent cortége, performing his high behests, follow whithersoever he leads through space!

Here we reach the boundary which divides the known from the unknown. Steadily we have pursued the human mind as it has moved on in its grand researches of the universe of God. Time, and space, and number, and distance, have all been set at defiance. No limits have been sufficiently great to circumscribe its movement. For more than six thousand years, onward! has truly been the word. And here I might very well pause, and rest content in the exhibition of the absolute and actual triumphs of human genius; but as the rays of the rising sun penetrate the darkness of night, and scattering the gloom, dimly reveal the scenes of earth which are soon to be flooded with splendor, so the light of human knowledge breaks over the boundaries which divide the known from the unknown, and faintly reveals what yet lies far beyond in the dark profound.

Guided by this light, we shall ask your attention to one of the most sublime speculations to which the mind of man has ever risen. I refer to the supposed discovery of the great centre about which it is presumed the **myriads** of stars composing our mighty **Milky Way** are all revolving.

M. Maedler, the author of the recent investigations with reference to the *Central Sun*, has long been known to the astronomical world as the successor of M. Struve in the direction of the observatory at Dorpat. His computations of the orbitual movements of the double stars have given to him a deservedly high celebrity, and the great theory which he has propounded is only given to the world after a long and patient examination extending through seven years.

The extension of the law of gravitation to the fixed stars, now absolutely demonstrated in the revolutions of the binary systems, settles forever the fact, that in the grand association of stars composing our cluster, or, as we shall hereafter call it, our *astral system*, there must be a *centre of gravity*, as certainly as there is one to the solar system. In the organization of the solar system we find a central body of vast size, surrounded by small and subordinate satellites. Again, among the planets, we find their magnitude very great, when compared with the moons which circulate around them. Extending this analogy, early astronomers conceived that this principle of a great central preponderating globe would, in all probability, obtain in all the higher orders of physical organization.

This idea, apparently so well founded, was entirely destroyed by the discovery of the binary stars. Here we find the next higher organization above our solar system, but instead of finding in the bodies thus united a vast preponderance in magnitude of one over the other, there are many examples in which the two suns thus united by gravitation are, in all respects, equal. In many others the difference is only slight, yet in all these higher systems there must exist a common centre of gravity.

With the mind cleared, by these views, from all prejudice in favor of the necessary existence of some stupendous central globe, as far exceeding in magnitude the myriads of fixed stars by which it is surrounded as does the sun all the satellites of its system, we are prepared to inquire into the actual existence or non-existence of such a body.

Admitting its invisibility, either in consequence of its distance or non-luminous character, there are yet remaining the means, not only of detecting its existence, but of discovering its position in space. In case such a body exists, the stars located nearest to it will be most completely subjected to its influence, and will show their proximity by the swiftness of their motion. Since it is possible to penetrate space in every direction, in case the stars of any particular region were endowed with a more rapid motion than all others, these would not fail to be discovered. But no such rapid motions have ever been detected, and hence it is now fair to conclude that such motions do not exist, and consequently no vast central globe can ever be found, because there is no evidence that such a body has any locality in space.

The questions resolve itself, then, into a research for the common centre of gravity of all the stars composing our astral system, and the data for such an examination must be found in the direction of the solar motion, and in that of the proper motion of the fixed stars. Difficult as this research undoubtedly is, Maedler's sagacity detected various guides which limited his more minute examinations to a comparatively small portion of the heavens. Since our great astral system has been shown to take the form of a layer or stratum whose thickness is small compared with its extent, we cannot fail to perceive that the centre of gravity of a mass of stars thus arranged must be found somewhere within the limits of the Milky Way, when seen by an eye located not very distant from the centre. But it is seen that our sun does not occupy the absolute centre of this stratum. In case it did, then would the bright circle of the Milky Way divide the heavens into two equal hemispheres. Since there is a manifest difference between the two parts into which the heavens is divided, the smaller portion will be the more distant from us, and in this smaller part we must look for the central point. But, from the soundings of both the Herschels, it is certain that our sun lies nearer the southern half of the Milky Way than the northern. Hence, in our researches for the centre of gravity, we may confine our examinations to the northern half of the smaller of the two parts into which the Milky Way divides the heavens.

One more approximation may be made. If we knew that our sun, in its presumed revolution about this great centre, described a circle, and if we knew the plane of this circle, and the direction in which the sun was now moving, a line drawn in that plane from the sun, and in a direction perpendicular to its line of motion, would pass directly through the centre about which it is revolving, and would point us directly to it. Now the direction of the sun's motion is alone determined; but since the centre of gravity must be found somewhere in a line perpendicular to the direction, we must give to this perpendicular all possible positions in space, which will cause it to cut from the celestial sphere the circumference of a great circle, within which the centre of gravity must be found.—These limiting considerations brought the distinguished astronomer to a region of the heavens in and about the constellation Taurus.

Here the examination took a more definite and more strictly scientific form. The proper motion of the stars in this region could be anticipated and known, at least in character and direction. The great centre would probably be located within the limits of some rich cluster. All the stars composing this cluster as well as those within 20° or 30°, would appear to move in the same direction. Those immediately proximate to the central sun or star would appear to move with the same velocity due to that star, and the entire group would sweep, apparently, through space without parting company.

Having, by such like considerations, narrowed down the limits of research, Maedler commenced his individual examinations. Among other objects subjected to rigid scrutiny, was the brilliant star Aldebaran, in the eye of the Bull. This being the brightest star in this region, and being, moreover, in the midst of a group of smaller stars, seemed, in the outset, to fulfill some of the conditions required of the central sun. But a more rigid examination proved conclusively that this star could not occupy the centre. Its own proper motion far exceeded that of the surrounding stars, and demonstrated its near proximity to our own system, and its mere optical connection with the stars surrounding it.

Thus did this great astronomer move from point to point, from star to star, subjecting each successively to the severest tests, until, finally, a point was found, a star was discovered, fulfilling, in the most remarkable manner, all the requisitions demanded by the nature of the problem. All are familiar with the beautiful little cluster, called the Pleiades, or seven stars. Clustered around the brilliant star Alcyone, which occupies the optical centre of the group, the telescope shows fourteen conspicuous stars. The proper motions of all these have been determined with great exactitude. These are *all in the same direction*, and are all nearly equal to each other; and, what is still more important, the mean of their proper motions differs from that of the central star, Alcyone, by only one thousandth of a second of arc in right ascension, and by two-thousanth's of a second in inclination.—Here, then, is a magnificent group of suns, either actually allied together, and sweeping in company through space, or else they compose a cluster so situated as to be affected by the same apparent motion produced by the sun's progressive motion through the celestial regions.

But an extension of the limits of research around Alcyone exhibits the wonderful truth, that out of one hundred and ten stars within 15° of this centre, there are sixty moving south, or in accordance with the hypothesis that Alcyone is the centre, forty-nine exhibiting no well defined motion, and only one single individual which appears to move contrary to the computed direction !

It is impossible, here, to do justice to the profound and elaborate investigations of the learned author of this great speculation. Assuming Alcyone as the grand centre of the millions of stars composing our astral system, and the direction of the sun's motion, as determined by Argelander and Struve, he investigates the consequent movements of all the stars in every quarter of the heavens. Just where the swiftest motions should be found, there they actually exist, either demonstrating the truth of the theory, or exhibiting the most remarkable and incredible coincidences. We shall not pursue the research. After a profound examination, Maedler reaches the conclusion *that Alcyone, the principal star in the group of the Pleiades, now occupies the centre of gravity, and is at present the sun about which the universe of stars composing our astral system are all revolving.*

Here, then, we stand on the confines of the unknown. One mighty effort has thus been made to bring beauty and order out of the chaos of motion which has hitherto distinguished the stars of heaven. Once the planets, freed from law, darted through space, or relaxing their speed actually turned back on their unknown routes. Chaos reigned among these flying globes until the mind, rising by the efforts of its own genius, reached the grand centre of the planetary orbs, and lo! confusion ceased, and harmony and beauty held their sway among these circling worlds. The same daring human genius which, sweeping across the interplanetary spaces, finally reached the controlling centre of our own great system, has now boldly plunged into the depths of space, has swept across the interstellar spaces, and roaming from star to star, from sun to sun, from system to system, looks out upon the universe of stars, and seeks that point from whence these millions of sweeping suns shall exhibit that grand and magnificent harmony which doubtless reigns throughout the vast empire of Jehovah.

We are too apt to turn away from the first efforts to resolve these mighty problems. How were the doctrines of Newton received? How much regard was paid to Herschel's grand theory of the solar motion? And yet how triumphantly have these great theories been established. But do you inquire if there be any possibility of proving or disproving the doctrines of Maedler? The answer is simple. Should the time ever come when the direction of the solar motion shall be sensibly changed, in consequence of its curvilinear character, then will the plane in which this movement lies be revealed, and then the centre about which the revolution is performed must be made known, at least in direction. Should the line reaching towards this grand centre pass through Alcyone this added to all the other evidences, will fix forever the question of its central position. We know not when this great question may be settled, but judging from the triumphs which have marked the career of human genius hitherto, we do not dare to doubt of the final result.

Admitting the truth of Maedler's theory, we are led to some of the most astonishing results. The known parallax of certain fixed stars gives to us an approximate value of the parallax of Alcyone, and reveals to us the distance of the grand centre. Such is the enormous interval separating the sun from the central star about which it performs its mighty revolution, that the light from Alcyone requires a period of 537 years to traverse the distance! And if we are to rely on the angular motion of the sun and system, as already determined, at the end of 18,200,000 years, this great luminary, with all its planets, satellites, and comets, will have completed *one* revolution around its grand centre!

Look out to-night on the brilliant constellations which crowd the heavens. Mark the configurations of these stars. Five thousand years ago the Chaldean shepherd gazed on the same bright groups.—Two thousand years have rolled away since the Greek philosopher pronounced the eternity of the heavens, and pointed to the ever-during configuration of the stars as proof positive of his assertion. But a time will come when not a constellation now blazing in the bright concave above us shall remain. Slowly, indeed, do these fingers on the dial of heaven mark the progress of time. A thousand years may roll away with scarce a perceptible change;—even a million of years may pass without effacing all traces of the groupings which now exist; but that eye which shall behold the universe of the fixed stars when ten millions of years shall have silently rolled away, will search in vain for the constellations which now beautify and adorn our nocturnal heavens. Should God permit, the stars may be there, but no trace of their former relative positions will be found!

Here I must close. The intellectual power of man, as exhibited in his wonderful achievements among the planetary and stellar worlds, has thus far been our single object. I have neither turned to the right hand nor to the left. Commencing with the first mute gaze bestowed upon the heavens, and with the curiosity awakened in that hour of admiration and wonder, we have attempted to follow rapidly the career of the human mind, through the long lapse of six thousand years. What a change has this period wrought. Go backward in imagination to the plains of Shinar, and stand beside the shepherd astronomer as he vainly attempts to grasp the mysteries of the waxing and waning moon, and then enter the sacred precincts of yonder temple devoted to the science of the stars. —Look over its magnificent machinery; examine its space-annihilating instruments, and ask the sentinel who now keeps his unbroken vigil, the nature of his investigations.

Moon, and planet, and sun, and system, are left behind. His researches are now within a sphere to whose confines the eagle glance of the Chaldean never reached. Periods, and distances, and masses, and motions, are all familiar to him, and could the man who gazed and pondered six thousand years ago stand beside the man who now fills his place, and listen to his teachings, he would listen with awe, inspired by the revelations of an angel of God. But where does the human mind now stand? Great as are its achievements, profoundly as it has penetrated the mysteries of creation, what has been done is but an infinitesimal portion of what remains to be done.

But the examinations of the past inspire the highest hopes for the future. The movement is one constantly accelerating and expanding. Look at what has been done during the last three hundred years, and answer me to what point will human genius ascend, before the same period shall again roll away? But in our admiration for that genius which has been able to reveal the mysteries of the universe, let us not forget the homage due to Him who created, and by the might of his power sustains all things.

At some future time, I hope to be permitted to direct your attention to this branch of the subject. If there be anything which can lead the mind upward to the Omnipotent Ruler of the universe, and give to it an approximate knowledge of His incomprehensible attributes, it is to be found in the grandeur and beauty of His works.

If you would know his *glory*, examine the interminable range of suns and systems which crowd the Milky Way.—Multiply the hundred millions of stars which belong to our own "island universe" by the thousands of these astral systems that exist in space, within the range of human vision, and then you may form some idea of the infinitude of his kingdom; for lo! these are but a part of his ways. Examine the scale on which the universe is built.—Comprehend, if you can, the vast dimensions of our sun. —Stretch outward through his system, from planet to planet, and circumscribe the whole within the immense circumference of Neptune's orbit. This is but a single unit out of the myriads of similar systems. Take the wings of light, and flash with impetuous speed day and night, and month, and year, till youth shall wear away, and middle age is gone, and the extremest limit of human life has been attained;—count every pulse, and at each speed on your way a hundred thousand miles; and when a hundred years have rolled by, look out, and behold! the thronging millions of blazing suns are still around you, each separated from the other by such a distance that in this journey of a century you have only left half a score behind you.

Would you gather some idea of the *eternity* past of God's existence, go to the astronomer, and bid him lead you with him in one of his walks through space; and as he sweeps outward from object to object, from universe to universe, remember that the light from those filmy stains on the deep pure blue heaven, now falling on your eye, has been traversing space for a million of years. Would you gather some knowledge of the *omnipotence* of God, weigh the earth on which we dwell, then count the millions of its inhabitants that have come and gone for the last six thousand years. Unite their strength into one arm, and test its power in an effort to move this earth. It could not stir it a single foot in a thousand years; and yet under the omnipotent hand of God, not a minute passes that it does not fly for more than a thousand miles. But this is a mere atom;—the most insignificant point among his innumerable worlds. At his bidding, every planet, and satellite and comet, and the sun himself, fly onward in their appointed courses. His single arm guides the millions of sweeping suns, and around His throne circles the great constellation of unnumbered universes.

Would you comprehend the idea of the *omniscience* of God, remember that the highest pinnacle of knowledge reached by the whole human race, by the combined efforts of its brightest intellects, has enabled the astronomer to compute approximately the perturbations of the planetary worlds. He has predicted roughly the return of half a score of comets. But God has computed the mutual perturbations of millions of suns, and planets, and comets, and worlds, without number, through the ages that are passed and throughout the ages which are yet to come, not approximately, but with perfect and absolute precision. The universe is in motion,—system rising above system cluster above cluster, nebula above nebula,—all majestically sweeping around under the providence of God, who alone knows the end from the beginning, and before whose glory and power all intelligent beings, whether in heaven or on earth, should bow with humility and awe.

Would you gain some idea of the *wisdom* of God, look to the admirable adjustments of the magnificent retinue of planets and satellites which sweep around the sun. Every globe has been weighed and poised, every orbit has been measured and bent to its beautiful form. All is changing, but the laws fixed by the wisdom of God, though they permit the rocking to and fro of the system, never introduce disorder, or lead to destruction. All is perfect and harmonious, and the music of the spheres that burn and roll around our sun, is echoed by that of ten millions of moving worlds, that sing and shine around the bright suns that reign above.

If overwhelmed with the grandeur and majesty of the universe of God, we are led to exclaim with the Hebrew poet king,—" When I consider thy heavens, the work of thy fingers, the moon and the stars which thou hast ordained, what is man that thou art mindful of him? and the son of man, that thou visitests him?" If fearful that the eye of God may overlook us in the immensity of his kingdom, we have only to call to mind that other passage, "Yet thou hast made him but a little lower than the angels, and hast crowned him with glory and honor. Thou madest him to have dominion over all the works of thy hand; thou hast put all things under his feet." Such are the teachings of the word, and such are the lessons of the works of God.

www.ingramcontent.com/pod-product-compliance
Lightning Source LLC
Chambersburg PA
CBHW020249170426
43202CB00008B/289